I0426770

Nuclear Cooperation with India:
New Challenges, New Opportunities

Edited by
Wade L. Huntley and Karthika Sasikumar

Simons Centre for Disarmament
and Non-Proliferation Research

Vancouver, BC, Canada

Nuclear Cooperation with India:
New Challenges, New Opportunities

Edited by Wade L. Huntley and Karthika Sasikumar

© 2006: The Simons Centre for Disarmament and Non-Proliferation Research (4)

All rights reserved. Small excerpts of this publication may be reproduced, downloaded, disseminated and transferred for academic/scholarly purposes without permission as long as express attribution is given to the publication source and to the Simons Centre for Disarmament and Non-proliferation Research at the University of British Columbia as publisher. Reproduction, storage or transmission of substantial portions of this publication requires the prior written permission of the Simons Centre. The unaltered complete electronic (PDF) version of the publication may be freely copied and disseminated without prior notification.

Simons Centre for Disarmament and Non-Proliferation Research
Liu Institute for Global Issues, University of British Columbia
6476 NW Marine Drive
Vancouver, BC V6T 1Z2 Canada
Tel: 604-822-0483
Fax: 604-822-9261
Email: simons.centre@ubc.ca
Website: www.ligi.ubc.ca

Cover Images: Top, US President George W. Bush and Indian Prime Minister Manmohan Singh in New Delhi, March 2, 2006 (Photograph: Gurinder Osan / AP); Bottom, India's CIRUS nuclear reactor at Trombay, Mumbai, provided by Canada in 1956.
Cover design by Wade L. Huntley

CONTENTS

Introduction

Presentations

Discussion

Conclusion

Background Papers

Bibliographies and Biographies

Abbreviations

ABM	Anti-Ballistic Missile
AECL	Atomic Energy Canada Limited
BJP	Bharatiya Janata Party
CANDU	Canadian Deuterium Uranium
CIRUS	Canada India Reactor US
COG	CANDU Owners Group
CTBT	Comprehensive Test Ban Treaty
DAE	Department of Atomic Energy (India)
DFAIT	Department of Foreign Affairs and International Trade
ECIC	Exchange Credit Insurance Company
EDC	Export Development Canada
EU	European Union
FMCT	Fissile Materials Control Treaty
GDP	Gross Domestic Product
IAEA	International Atomic Energy Agency
MTCR	Missile Technology Control Regime
MW	Megawatts
NNPA	Nuclear Nonproliferation Act (US)
NNWS	Non-nuclear Weapons State
NPT	Nuclear Non-proliferation Treaty
NSG	Nuclear Suppliers' Group
NWS	Nuclear Weapons State
RAPS/RAPP	Rajasthan Atomic Power Station/Plant
UK	United Kingdom
UN	United Nations
UNSC	United Nations Security Council
US	United States
USSR	Union of Soviet Socialist Republics
VHP	Vishwa Hindu Parishad
WANO	World Association of Nuclear Operators
WMD	Weapons of Mass Destruction
WW II	World War Two

Introduction

Nuclear Cooperation with India: An Introduction

Wade L. Huntley

Introduction

On 18 July 2005, US President George Bush and Indian Prime Minister Manmohan Singh announced a bold agreement to restore US nuclear cooperation with India. The UK, France and other countries soon followed this lead. On 26 September 2005, Canada announced a new initiative seeking engagement with India but not yet lifting existing nuclear trade restrictions.

New nuclear cooperation with India is controversial: it has been both applauded and critiqued in the US and Canada, elsewhere in the world, and in India itself. Some suggest the deal went too far; others, not far enough. Certainly, the matters are complex and multifaceted. The long-term implications may not be fully known for years to come.

On 22 November 2005 the Simons Centre for Non-Proliferation and Disarmament Research convened a one-day conference, "Nuclear Cooperation with India," to seek new insights into these issues. The conference gathered a diverse group of specialists to discuss the political and technical consequences of the recent initiatives. The presentations and discussion considered the merits of the specific terms of the US-India arrangement, Canadian policy responses, the direct impact on the NPT regime, and the potential broader consequences for non-proliferation efforts worldwide. This volume documents the proceedings of the conference and includes background papers and suggestions for further reading.

The Setting

India has had a complex relationship to nuclear non-proliferation since the dawn of the nuclear age. Early in the Cold War period, India was a leader in the Non-Aligned Movement's unambiguous call for complete global nuclear disarmament. But India pursued early opportunities to develop a full-scale civilian nuclear energy program, including acquisition of nuclear reactors and fuel reprocessing technologies from Canada and the US.

In 1970, the NPT came into force. This treaty enshrined the goal of global nuclear disarmament, but also allowed continued possession of nuclear weapons by the five countries that had tested a nuclear explosive device prior to 1 January 1967. India was not one of these countries, and could therefore only join the treaty as a non-nuclear state.

Four years later, India conducted a 'peaceful nuclear explosion' using knowledge and materials developed with the benefit of imported civilian technologies. This alienated many in the international community, especially Canada and the US, which had provided India with nuclear technology under the expectation it would be used only for civilian power generation. India, for its part, maintained a rhetorical commitment to global disarmament but refused to join the NPT as a non-nuclear state. This tension has defined India's relationship with the global non-proliferation regime ever since.

In 1998, in the context of the growing momentum toward a global CTBT, India conducted a series of nuclear weapons tests, establishing itself as a *de facto* NWS. When Pakistan quickly followed with nuclear tests of its own, many feared the onset of a new era of nuclear insecurity in South Asia.

Shortly after the South Asian nuclear tests I wrote an article exploring four possible scenarios for future developments (see Figure 1, below) (Huntley 1999). Three of these scenarios were dire, involving a nuclear arms race and/or conflict between India and China. But the fourth scenario, in which such ominous consequences would not emerge, presented a dilemma. In this scenario, which I called 'Nuclear Peace,' relative stability would reinforce perceptions of the strategic and political efficacy of nuclear weapons and, ironically, pose the greatest challenge to the non-proliferation regime.

This fourth scenario conveys much of what has actually transpired in the years since India's and Pakistan's nuclear tests. India-Pakistan relations have not exploded, figuratively or literally. Indeed, while some regarded the Kargil war and the 2001-02 border standoff as playing dangerously with nuclear matches, others saw nuclear deterrence in action, inducing caution and restraining what might otherwise have escalated into much wider conflicts. Subsequently, India-Pakistan relations have progressed, in some ways remarkably. India-China relations have also seen significant improvement since 1998. Accordingly, many Indian advocates of nuclear weapons feel events have validated their judgments. As India's emergence as a NWS has become an established fact, India has also shifted its non-

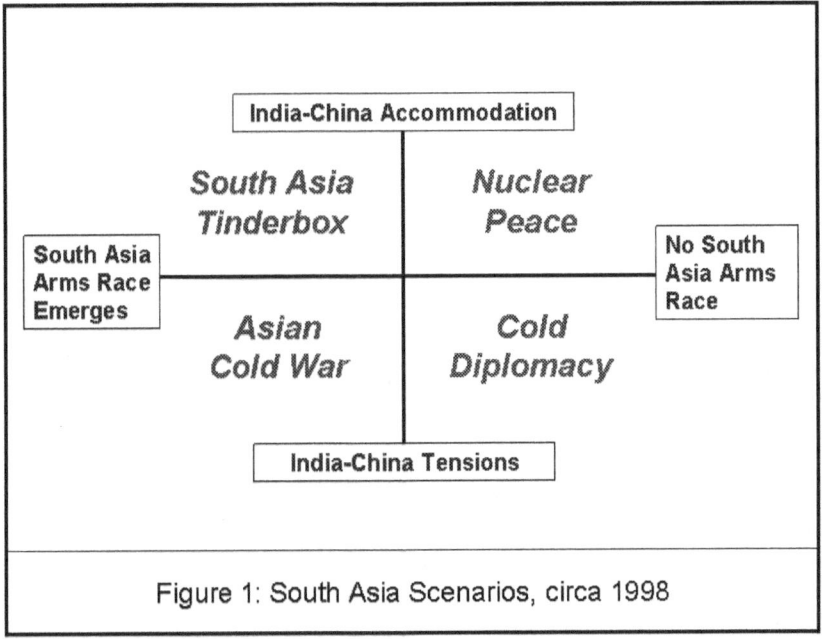

Figure 1: South Asia Scenarios, circa 1998

proliferation policies, adopted a 'minimal deterrence' nuclear doctrine, and sought international legitimacy as a 'responsible' nuclear power.

In the meantime, the NPT continued to gain non-nuclear adherents and is now an almost globally inclusive regime – India and Pakistan are two of only three countries not to have acceded to the treaty. (The third is Israel, also now widely thought to possess nuclear weapons. North Korea, suspected of developing nuclear weapons, in 2003 became the only state to have withdrawn from the NPT.) The NPT has emerged as a centerpiece of a wider non-proliferation regime including other agreements and associations, such as the NSG, which seeks to regulate exports of nuclear power technologies that could contribute to nuclear weapons development. Any cooperation with India on nuclear matters, many have felt, would serve to legitimate its nuclear weapons acquisition and erode the NPT regime.

Admitting India to the NPT as a NWS would undermine the treaty's foundation. But the inadmissibility of India as a NWS directly conflicts with India's declared intention to be a full-fledged nuclear power. This tension has grown with the ongoing development of India's economic and political roles in South Asia and the world. Following the terrorist attacks of 11 September 2001, new attention to the threat of proliferation of nuclear technologies and materials to 'rogue states' and terrorists increased global interest in gaining Indian ascent to the kind of non-proliferation commitments NPT states accept.

This growing tension finally snapped when President Bush and Prime Minister Singh announced the agreement to restore US nuclear cooperation with India. While the course of future events is uncertain, it is likely that both the NPT's and India's roles in global non-proliferation efforts will be permanently altered.

The Conference

Conference presentations delved into the implications of new nuclear cooperation with India from several directions. Following is a brief overview of the presentations by the eight invited panellists.

Ernie Regehr (Project Ploughshares, Canada) observed that India, as a major power, can do much to either contribute to or frustrate global non-proliferation efforts. However, the US-India deal does little to expand India's efforts beyond its current practices. The central non-proliferation challenge with regard to India, he maintained, is to encourage its support of the non-proliferation regime – including strict adherence to international restrictions on the transfer of nuclear materials and technology – without explicitly or even implicitly accepting India as a NWS and thus ignoring its own past violations of non-proliferation norms. Regehr emphasized that, at a minimum, changes in policy toward India must be made within a multilateral context and in strict compliance with global standards, without resorting to selective non-proliferation and exceptionalism.

Seema Gahlaut (Center for International Trade and Security, University of Georgia, US) contended directly with critics who claim that new nuclear cooperation with India will undermine the moral authority of the NPT, lead to nuclear reversals and set an avoidable precedent of discriminating in favour of a 'proliferator' state. She asserted that most such arguments rely on either moral or narrowly-based legal reasoning that has little empirical basis and breaks down under closer examination. Gahlaut argued that focusing exclusively on the sanctity of the NPT forecloses pragmatic policy innovations without presenting viable alternatives. Consequently, the international community risks missing vital opportunities to engage India in global non-proliferation efforts and unintentionally strengthening the regime's foes within the country.

M.V. Ramana (Centre for Interdisciplinary Studies in Environment and Development, Institute for Social and Economic Change, Bangalore, India) argued that India needs neither nuclear energy for its development nor nuclear weapons for its security. Both India's nuclear 'hawks' and its nuclear 'nationalists' believe in the past and future success of India's nuclear energy program, despite a long-standing record of cost overruns and failures. Both groups also subscribe to the security benefits of an Indian nuclear arsenal despite strong strategic, legal and moral indications to the contrary. Ramana maintained that new nuclear cooperation with India would enable the country to import fuel for civilian reactors, freeing up domestic uranium supplies for rapid expansion and diversification of its nuclear weapons arsenal, including possible creation of both smaller-yield fission weapons and larger-yield thermonuclear weapons.

Neil Joeck (Center for Global Security Research, Lawrence Livermore National Laboratory, US) acknowledged as reasonable many concerns over the non-proliferation impacts of new nuclear cooperation with India, but argued that the NPT has been but one element of a long history of non-proliferation efforts characterized by adaptation and change. While many policies continue to work well, others have proven limited or have weakened over time as new contingencies have emerged, necessitating an ongoing search for new ways to achieve non-proliferation and security objectives. Joeck contended that the new US nuclear cooperation agreement with India represents such an innovation, aimed at enlisting India as an ally in global non-proliferation efforts and exemplifying a layered approach that combines new initiatives with existing instruments such as the NPT. Thus, forging new partnerships with India provides a means to increase global security while adapting to new conditions.

Ross Neil (Former Nuclear Non-proliferation Officer, Canadian Nuclear Safety Commission) surveyed Canadian government policies, procedures and actions vis-à-vis its nuclear dealings with India (and Pakistan) in the years leading up to the NPT entry-into-force in 1970 and India's first nuclear explosive test in 1974, and extended his analysis to Canada's options in the post-1974 period. As Canada's nuclear non-proliferation policy experienced a coming of age, the domestic political context constrained implementation of international nuclear non-proliferation norms and ongoing events compelled the adaptation of non-proliferation policy. Neil raised the question of whether the relative rigidity of Canadian policy since the mid 1970s, combined with the difficulty of bringing India into the NPT as a non-nuclear state and the general lack of progress toward general nuclear disarmament, have resulted in significant lost opportunities for Canada to advance its non-proliferation objectives in other ways.

Ron Stansfield (Nuclear Arms Control and Disarmament Division, Foreign Affairs Canada) observed that the US-India agreement has significantly altered the international security landscape, and noted that US re-engagement of India creates a number of challenges for Canadian policy in light of Canada's own difficult history with India on nuclear issues. He cautioned that the rehabilitation of India in the nuclear field poses a number of conundrums to traditional nuclear non-proliferation and disarmament approaches, and that implementation of the US-India deal requires legislative changes that the US Congress may resist. Stansfield observed that Canada seeks strengthened economic and political ties with India, pointing to the emphasis on enhancing the bilateral relationship in the Martin Government's 2005 International Policy Statement. Canada's 26 September 2005 statement on nuclear engagement with India sought to credit India's non-proliferation efforts, but did not relax existing nuclear trade restrictions or mark a substantive change in current Canadian nuclear non-proliferation policy. Stansfield concluded that if the time has come for new non-proliferation efforts less centered on the NPT, managing the consequences of these initiatives for the existing regime remains paramount.

S. Paul Kapur (Center for International Security and Cooperation, Stanford University, US) judged that new nuclear cooperation with India is likely to have mixed impacts depending on the issue area. Economically, the boost to India's civilian nuclear energy production would partly substitute for inefficient and environmentally degrading fossil fuel consumption as energy needs grow, despite continuing problems in the civilian program. Impacts in terms of diplomacy and security are more ambivalent: *de facto* US recognition of India as a NWS will create pressures on India to accommodate US interests vis-à-vis Iran and China that might not be readily forthcoming, and the potential boost to India's nuclear and conventional military strength could increase or decrease stability in regional relations. Regarding non-proliferation efforts, Kapur asserted that the US-India deal would likely damage the NPT by undermining nuclear supplier regulations, possibly encouraging current NPT signatories to leave the treaty, and implicitly jettisoning the NPT's foundational precept that all nuclear proliferation is equally undesirable. These costs could outweigh the benefits of incorporating India into the regime.

T.V. Paul (McGill University, Canada) shared the opinion that new nuclear cooperation with India has mixed implications. On the one hand, it does generate questions about equity, fairness, and the treatment of different new nuclear states and non-nuclear states by the US and other nuclear "haves." But, he asserted, the US-India nuclear accord will probably have a minimal impact on the regime and is unlikely to lead to additional states acquiring nuclear weapons, because states tend to acquire or give up nuclear weapons on the basis of nearby situational factors rather than distant examples. Paul contended that India is essentially a status quo power that has restrained itself from spreading weapons or materials to other states, unlike other revisionist states that are the most notable proliferation concerns today. A major problem with the NPT is that its inflexibility to shifting global power dynamics makes it fragile, and it could contribute to tension in the international system if it cannot find a way to accommodate rising powers. Paul concludes that incorporating India into the global nuclear order would promote international stability and help restrict nuclear proliferation.

The rest of this publication includes fuller texts of each presentation, a summary of the ensuing discussions, and a concluding essay that highlights the most salient points of divergence at the conference and draws out the key questions and uncertainties for the future. The volume also includes supplementary background material and a current bibliography.

The editors' hope is that this publication will provide a timely survey of the complexities of the issues involved in initiating new civilian nuclear cooperation with India, and contribute to ongoing debates over present and future policy options.

Presentations

An Exception to the Rule? The US-India Nuclear Cooperation Deal and Non-proliferation Risks

Ernie Regehr

The debate over the proposed US-India deal reproduces with discouraging fealty the fundamental non-proliferation divide between states with nuclear weapons and those without. NWS are certainly expected to make important commitments to disarmament, but in practice it is assumed they will temper implementation of those commitments in deference to their special national interests and strategic objectives. NNWS, on the other hand, have clear, measurable obligations, the implementation of which is not optional.

The interests of states with nuclear weapons seem by definition to enjoy a gravitas that demands deference, and it is on full display in the proposed US-India deal. The deal calls on India to continue doing what it is already obliged to do, and what to its credit it generally does – that is, exercise careful control over nuclear trade – but after that the demands on India are modified at every turn by its own particular interests:

- India is asked to separate the civilian and military elements of its nuclear program, but according to its own preferences

- Civilian facilities are to be voluntarily placed under IAEA safeguards, arrangements that can be reversed if circumstances are deemed to warrant it

- The Additional Protocol requirement is a courtesy without substance inasmuch as the key provision of the protocol, that India provide access to all nuclear facilities, does not apply. In this case, like the NWS, India still decides which facilities are covered and which are excluded

- India is called on to continue its testing moratorium, but is under no obligation to enshrine it as a legal commitment by signing and ratifying the CTBT

- India gives assurances that it will support efforts toward the FMCT, but in the meantime accepts no obligation to stop producing fissile material for weapons.

In one sense the debate over the US-India deal seems simply to be between those who believe in a rules-based non-proliferation regime and those who, while sharing that belief, allow exceptions to the rules – with the US as the general exception and, in this case, India as the specific exception.

That is not to say that those in the no-exceptions camp do not also recognize that the current rules are ineffective in addressing India's situation, or that rules may need to change. But they do insist that if India does in fact not fit existing rules, it does not follow that it then deserves an exemption from them. If the rules are inadequate, then they need to be changed through multilateral agreement to meet non-proliferation requirements in the kinds of circumstances that India exemplifies, applying to all, not just to India and the US.

It is obvious that the international community faces a genuine dilemma regarding India. A major economic, industrial, military, and democratic power, India can do much to contribute to or frustrate global non-proliferation efforts. Therefore, the question is: how can India be mobilized in support of nuclear non-proliferation – including strict adherence to international restrictions on the transfer of nuclear materials and avoidance of regional nuclear weapons competition – without exempting India from international law and global norms or standards.

In late 2005 the international community once again, by now in largely *pro forma* fashion, set out what is required of India. The UN First Committee called upon India, along with Pakistan and Israel, to join the NPT as a NNWS, "promptly and without condition." That position has near universal support – only three states voted against the resolution (A/C.1/60/L.4), namely India, Israel, and Pakistan. The US, UK, and France abstained, but not, at least in the case of the latter two, because of the reference to the three NPT holdouts.

In 2000 all the states parties to the NPT, "call[ed] upon those remaining States not party to the Treaty to accede to it, thereby accepting an international legally binding commitment not to acquire nuclear weapons or nuclear explosive devices and to accept IAEA safeguards on all their nuclear activities." The 2000 NPT Review Conference explicitly declared that the nuclear tests by India and Pakistan "do not in any way confer a nuclear-weapon-State status" on them (Anon 2000).

That is the current global standard. It was unilaterally set aside when President Bush announced that under Washington's move toward "full civil nuclear energy cooperation and trade," India would be exempted from the international community's requirement of NPT membership as a NNWS and instead "should receive the benefits and accept all the responsibilities of the world's leading states with advanced nuclear technology" (Office of the Press Secretary 2005) – in other words, be regarded as a NWS.

What are the proliferation risks inherent in this Indian exceptionalism?

Exceptionalism and selective non-proliferation are flip sides of the same coin. They represent what US Democratic Representative Ed Markey called "picking and choosing when to pay attention to the existing non-proliferation treaties" (Markey 2005). So the question is, how does this "picking and choosing" policy of selective exemptions from non-proliferation norms – a strategy that ultimately worries less about the spread of nuclear weapons than about who gets them – risk further vertical and horizontal nuclear proliferation?

While it is not possible to conclude with any certainty what the non-proliferation *consequences* of the US-India nuclear cooperation agreement might be, we can do what much of the commentary on the agreement has already done, and that is to try to understand the proliferation *risks* and explore ways to mitigate them. The following suggests four basic kinds of proliferation risk.

Enhancing the Political Role of Nuclear Weapons

The demand for nuclear weapons could be increased by a deal that portrays nuclear weapons as sources of status and power. "US acceptance of India as a nuclear-weapon state gives weight to the notion that nuclear weapons enhance a country's status and power" (McGoldrick, Bengelsdorf and Scheinman 2005).

Undermining Confidence in the Non-proliferation Regime

Trust in the non-proliferation regime stands to be weakened most in those technologically capable states that are especially critical to the continuation of the basic NPT bargain – that is, foreswearing nuclear weapons in exchange for access to nuclear technology and materials for peaceful uses and a commitment to disarmament: "Presumably, states that have relinquished nuclear weapons to become members of the NPT, such as South Africa and the Ukraine, strongly resent the deal. States like Argentina, Brazil and Egypt, which have explored a nuclear weapons option, but voluntarily chosen NPT membership instead, probably resent it too; the states which explicitly based their NPT membership on the premise that the international community would not recognize any additional NWS likewise" (Lodgaard nd).

Confidence in public institutions and agreements is perhaps the most precious and precarious ingredient of a successful rules-based order. For states to forego options and apparent advantages in favour of the collective will requires the confidence that others will also defer to the collective will and be disciplined by agreed rules.

Precedence

The deal opens the door to the possibility of other threshold states making the transition to nuclear acquisition and then insisting that they be treated as India is. The exceptionalism extended to India will necessarily extend to Israel and Pakistan. The logic of India as a special case cannot realistically or logically be confined to India. Any exception granted to India will certainly be claimed as the new standard by Iran. Its formal claim as a signatory to the NPT will not be to the same 'right' as India to acquire nuclear weapons. The issue for Iran is enriching and reprocessing. "In the near term, it will make it more difficult to deal with proliferation challenges such as Iran. Already the Iranians are winning support internationally by asking why they, as an NPT party, should give up their right to an enrichment capability while India, which rejected the NPT, is being offered nuclear cooperation" (Einhorn 2005).

Vertical proliferation

A deal that permits India to continue production of fissile material for weapons purposes while, at the same time, giving it full access to foreign uranium is a formula for unrestricted expansion of India's nuclear arsenal. Perkovich points out that this may well be a deliberate effort to facilitate an expanding Indian arsenal and that the failure to make a moratorium on fissile material production a part of the US-India deal "flows directly from the administration's priority of balancing Chinese military power" (Perkovich 2005). The result could obviously be a regional nuclear arms race.

There are two essential elements to mitigating the proliferation risks inherent in the US-India deal. The first is to make any rule changes part of a universal standard rather than a double standard. In addition, changes that are made should be through a process of collective consent that builds confidence in decision-making and that respects a broad global understanding of the common good with regard to nuclear proliferation.

The second is to toughen the deal itself. Two elements of such toughening would be to insist that India place a moratorium on fissile material production and that India sign the CTBT while continuing its moratorium on nuclear testing. The deal should also prohibit cooperation related to enrichment and reprocessing; indeed the deal should be seen as an opportunity to set a universal standard on control of the sensitive elements of the fuel cycle. In addition the deal should insist that safeguards on civil nuclear facilities apply in perpetuity. Finally, the deal should require that any changes in nuclear export practices be approved by the NSG.

A Critical Look at the Opposition to the US-India Agreement

Seema Gahlaut

The arguments against the Bush Administration's deal with India on civilian cooperation do not add up at all. The argument in favour of maintaining the purity of the NPT completely glosses over a few pertinent facts.

First, the argument that ignoring the NPT rules would lead to anarchy and chaos ignores the developments of the past 25 years when the NPT members themselves consistently broke every single rule of the NPT. Yet, NPT supporters find that the treaty remains worthy of support.

- Article I prohibits NWS from helping others acquire nuclear weapons. Yet 450-plus US nuclear weapons continue to be based in Europe, to be placed under the operational command of the host states in times of crises.

- Article II suggests that NNWS consider nuclear weapons irrelevant for their national security. Yet Germany, Japan, the Netherlands and Canada remain under the US nuclear umbrella. They have merely outsourced their nuclear defence, not done away with nuclear deterrence-based defence.

- NATO continues to expand its defensive doctrine to include the right to nuclear response when faced with chemical and biological attack.

- We now know that Chinese weapons designs traveled from Pakistan to Libya, and that European companies have long been the suppliers for A. Q. Khan's procurement efforts.

- Article VI violations are too numerous to be recounted, except to point out that in the absence of a hostile Soviet Union, and with European integration, there is no rationale for the UK and France to retain their nuclear arsenals, or for cooperation between these states and the US to upgrade the quality of their arsenals.

Second, the prediction of dire consequences - the domino effect in terms of nuclear reversals – has little basis in history. States do not build nuclear weapons – or even give them up – because other distant and/or non-threatening states have done so.

- Germany and Japan were made to give up their right to nuclear weapons due to their reprehensible activities during WWII. But even they were able to bargain with the international community: the clause about continued access to nuclear technology for peaceful purposes came into the NPT through their insistence. And they got a credible nuclear umbrella from the US, along with numerous economic benefits.

- Many other European states had secret nuclear ambitions that were thwarted by direct US pressure along with the incentive of a NATO umbrella, and an influx of light water reactors for energy purposes.

- Argentina and Brazil faced domestic change – overthrow of military juntas – and saw it as an opportunity to change their regional competition (not a rivalry with a history of wars) into mutual nuclear cooperation.

- South Africa's de-nuclearization was decided upon and completed by a discredited apartheid regime on its way out: the regime did not want the Black leadership to have a bomb.

- Belarus, Kazakhstan and Ukraine merely inherited arsenals from the Soviet Union. They gave up their right to these weapons in return for promises of substantial rewards from the West. Hardliners in Ukraine still feel cheated by this bargain.

Nuclear reversals among these "responsible" states are not likely to be a result of India's access to civilian nuclear technology, but from developments in their respective neighbourhoods. South Korea and Taiwan might respond to aggression, Japan might respond to an assertive North Korea or China, and Germany to a rejection of its UNSC bid.

This logic is evident even in the dynamic of states of proliferation concern: Egypt, Iran, Iraq, Libya, Syria, and Saudi Arabia have all been interested in nuclear weapons because of their hatred of Israel and their regional rivalries. If what is happening with India was a valid concern among these states, Libya would have drawn different conclusions about a post-1998 India instead of giving up its WMD programs.

Third, predictions about the NPT breaking up have been made before. That was the reason given for sanctioning India in 1978 and 1998 and for pressuring India to sign the CTBT. Yet, each time, it has had unintended detrimental consequences for the non-proliferation regime as a whole.

- The unilateral US nuclear and missile sanctions under the NNPA 1978 hardened the resolve of the Indian establishment. For instance, the denial of the Cray supercomputer by the Clinton administration spurred the development of the Indian supercomputer Param.

- The non-proliferation regime was lucky that India did not express its defiance of denials by selling technology to states of concern. The decision for restraint was based on a political commitment and national policy that such items are meant only to demonstrate Indian capabilities and to augment national defence.

- The CTBT euphoria in the non-proliferation community, especially the insertion of the Entry-into-force clause overriding India's objections, directly accelerated the decision to test nuclear weapons.

Fourth, the arguments that a rapprochement with India has to be rejected in order to 'save' the NPT totally ignore the undeniable truth that the NPT is like a constitution − broad in its sweep and entirely dependent upon subsequent interpretations for its implementation.

- The NPT's archaic and limited 'rules' had to be constantly shored up by a series of related agreements and institutions that try to close its gaping loopholes. These institutions do not bind themselves formally to the NPT, escaping the legal liabilities inherent in the so-called cornerstone of non-proliferation.

- The IAEA does not use NPT membership to allow/disallow technical cooperation. India is not merely a recipient but one of the largest contributors to the IAEA technical assistance programs from the developing world.

- The NSG seeks to complement the NPT but does not make NPT membership a sole criterion of participation. Moreover, it goes beyond NPT requirements to ask for fullscope safeguards and to control nuclear reactor equipment and related dual-use technologies. The NSG also maintains a 'Non-proliferation

Principle' whereby it reserves the right to deny controlled technologies to states regardless of their NPT compliance.

Discrimination, which these arguments seem to deplore when it is in favour of India, is at the heart of enforcing NPT norms. How else would the non-proliferation community have ensured that Japan is allowed to reprocess and stockpile plutonium when South Korea is not? Or that South Korean scientists can secretly enrich uranium to weapons grade, forge uranium metal from imported fertilizer, and reprocess plutonium – and yet not be reprimanded by the IAEA, but Iran is? Therefore, there is both precedent and logic for the non-proliferation community to not feel compelled to treat Pakistan – the locus of the 'nuclear Wal-Mart' – at par with India.

Finally, what alternate option would the non-proliferation community recommend to bring responsible non-members like India into the non-proliferation fold? The exhortations to India to do the right things – without providing any rewards – have been consistently ignored in the past. Yet, India, responding to the Bush administration's initiatives that included tangible rewards, harmonized its control list with those of the NSG and the MTCR in July 2005.

The NPT's success has depended critically on the "rule-breaking" side-deals cut by numerous US administrations - not unlike the current Bush administration. South Korea, Taiwan, Japan, Germany, Italy, and Ukraine did not just have a change of heart about nuclear weapons. The Clinton administration offered North Korea nuclear reactors in 1994 to bring it back into the NPT. Chinese nuclear and missile trade with Pakistan was downplayed in the 1990s so as to encourage China to reform its proliferation policies – and promote Western commercial interests.

The horror scenario of unrestrained Indian production of fissile material and warheads is based on the worst case predictions by the major critics of the agreement: no expert on India would assert that left to itself India is likely either to build nuclear weapons in indiscriminate quantities or aim its missiles at the US and Europe. India's nuclear program is much older and more extensive than those of Pakistan and China, yet there has been no Indian desperation to outpace either. One could argue, on the other hand, that opening up its vast civilian program to safeguarded nuclear cooperation would bring market efficiencies into its nuclear sector. Recent trajectories in India's high technology sectors show that the entry of the private sector attracts technical professionals away from government-owned defence enterprises, raising the pressure on the latter to become more open, transparent, and accountable.

Finally, it is short-sighted of the non-proliferation community to cast as inconsequential India's voluntary restraint in not proliferating beyond its borders because 'these are things India has been doing anyway.' The message that you have to become a non-proliferation prodigal (serious proliferator) before you can be welcomed back into the fold would be an extremely dangerous message to send to today's India. The era of idealism in India is decisively over. Now, the elite already makes clear linkages - publicly and privately - between using economic/technological capabilities and achievement of diplomatic and security goals, and vice versa. It would, therefore, not be a major leap of logic for hardliners to push for 'marketing' Indian dual-use technology following the example set by China and Pakistan.

Feeding the Nuclear Fire: Resuming Nuclear Cooperation with India

M. V. Ramana

There are two fundamental questions at the core of the Indo-US nuclear agreement. The first is whether India needs nuclear energy for its development. A good case can be made that it does not. The second is whether the country needs nuclear weapons if it wants to live in peace with the world. Many believe, with good reason, that it does not.

Much of the debate on the deal has been between what can be broadly called the nuclear hawks and the nuclear nationalists. The nuclear hawks believe India's nuclear program is a great success and more than able to take care of itself. They see the deal as imposing unnecessary constraints on the program and making more difficult the creation of the large nuclear arsenal, including thermonuclear weapons (hydrogen bombs), that they believe is essential for India to be a 'great power.'

Nuclear nationalists have the less ambitious, more traditional perspective that sees the nuclear program as a technological achievement necessary for India's economic and social development. They see the deal as offering a way to sustain and expand the energy program, while not unduly restricting the building of a 'minimum' nuclear weapons arsenal. The Indian government has embraced this view, as have many defenders of the deal. The Prime Minister laid it out most clearly to Parliament on 29 July 2005, saying: "Our nuclear program is unique. It encompasses the complete range of activities that characterize an advanced nuclear power... nuclear power has to play an increasing role in our electricity generation plans" and the deal offers a way where "our indigenous nuclear power program based on domestic resources and national technological capabilities would continue to grow." He made it clear that "there is nothing in the joint statement that amounts to limiting or inhibiting our strategic nuclear weapons program."

There are many problems with both views. The first is their shared belief in the success of India's nuclear energy program, and the need to continue with and expand this effort. They fail to recognize that the deal is actually testament to the long-standing, expensive, and large-scale failure of the DAE and its tremendous but largely hidden costs, in terms of health, safety, environment, and local democracy. The second belief shared by the nuclear hawks and the nuclear nationalists is that nuclear weapons are a source of security. This belief has been extensively debunked. Those who persist in this belief also ignore the essential moral and legal questions of what it means to have and be prepared to use nuclear weapons. The only difference between the two camps is on the character and size of the genocidal weapons they aspire to.

A History of Failure

The 1948 establishment of the AEC, which is now housed within the DAE, was framed by the rhetoric of indigenous national development. There was no progress until the UK offered the design details and enriched uranium fuel for the first Indian nuclear reactor, Apsara. In what was to become a pattern, the official announcements when the Apsara reactor went critical declared it a "purely indigenous affair" (Abraham 1997). Similarly, the CIRUS reactor, which provided the plutonium used in the 1974 nuclear test (and quite likely some used in the 1998 tests as well), was supplied by Canada while the heavy water used in it came from the US. Imported technology was not limited to reactors: many of India's nuclear scientists were made in America and elsewhere. Between 1955 and 1974, over 1100 Indian scientists were sent to train at various US facilities (Perkovich 1999).

Extensive foreign support ended only after the 1974 nuclear test. The international community, led by Canada and the US, incensed by India's use of plutonium from CIRUS that had been given to the country for purely peaceful purposes, cut off most material transfers. However, a little-advertised fact is that various nuclear facilities still procured components from abroad and foreign consultants continued to be hired. DAE personnel still had access to nuclear literature and participated in international conferences where technical details were freely discussed.

Even with all this help, DAE's failures were many and stark. In 1962, Homi Bhabha predicted that by 1987 nuclear energy would constitute 20,000 to 25,000 MW of installed electricity generation capacity (Hart 1983). His successor predicted that by 2000 there would be 43,500 MW of nuclear power (Sarabhai 1974). In 1984, the *Nuclear Power Profile* drawn up by the DAE suggested the more modest goal of 10,000 MW by 2000 (Ramachandran 2000). Today nuclear power amounts to only 3,400 MW, barely 3 per cent of India's installed electricity capacity. It is only with international help that the DAE can ever hope to achieve its latest promised goal of 20,000 MW by 2020.

Implications of the Agreement for Nuclear Energy in India

The other pressure driving this deal has been the DAE's extraordinary failure to manage its program. In its determination to build more and more reactors, something to show for all the money that it gets, the DAE has failed to take account of the need to fuel them. This was evident in the statement from an unnamed official to the British Broadcasting Corporation soon after the US-India deal was announced: "The truth is we were desperate. We have nuclear fuel to last only till the end of 2006. If this agreement had not come through we might as well have closed down our nuclear reactors and by extension our nuclear program" (Srivastava 2005). Aside from Tarapur I and II, all DAE reactors are fuelled using uranium from the Jaduguda region of Jharkhand. DAE may be only producing about 300 tonnes of uranium a year, falling short of the fuelling requirements. DAE has been able to continue to operate its reactors by using stockpiled uranium from earlier days when the nuclear capacity was much smaller. Estimates are that this stockpile would be exhausted by 2007. DAE has been desperately trying to open new uranium mines in the country. But it has been met with stiff public resistance everywhere (Dias 2005). This local resistance stems from the widely documented impacts of uranium mining and milling on public and occupational health.

If the deal goes through, DAE will be free to purchase uranium from the international market for its safeguarded reactors. This has some important consequences. The first of these is that it will reduce pressure on domestic uranium reserves. Since imported uranium will be much cheaper than Indian uranium, it may also marginally reduce the operating costs of Indian nuclear plants. Although the DAE hides its actual costs, there is little doubt that nuclear electricity is more expensive than other major sources of power in India (M. V. Ramana, D'Sa and Reddy 2005). The existing nuclear capacity and any increases should, therefore, not be considered a benefit.

Nuclear electricity is expensive and it would be far better to invest in other sources of power, as well as energy conservation measures. There are also important safety concerns associated with nuclear power. At least one of the DAE's nuclear reactors has come close to a major accident)(Chanda 1999) . Other facilities associated with the nuclear fuel cycle have also had accidents, though these have primarily affected workers within the plant. Apart from extreme accidents, there are many environmental and public health consequences associated with the many facilities that make up the nuclear complex (M.V. Ramana and Gadekar

2003). And, to cap it all, there is the so far unsolved problem of managing large amounts of radioactive waste.

How Many Bombs Are Enough?

Access to the international uranium market to fuel its power reactors will further free up domestic uranium for military uses. This may allow a significant expansion in India's nuclear weapons capabilities. There are several ways in which India could use its freed-up domestic uranium. It could choose to build a third reactor dedicated to making plutonium for its nuclear weapons. India could also start to make highly-enriched uranium for nuclear weapons - Pakistan has used such highly-enriched uranium, produced at Kahuta, for its weapons. Both paths, which need not be exclusive, would allow India to increase its fissile materials stockpile at a much faster rate. A third use for domestic uranium would be to supply the fuel for the nuclear submarine that has been under development since the 1970s (Rethinaraj 1998).

India can use both its current stockpile of weapons-grade plutonium and future production from CIRUS and Dhruva, and other future plutonium production reactors, to make nuclear weapons. The current stockpile is estimated to be perhaps 400-500 kilograms, sufficient for about 100 simple fission weapons. CIRUS and Dhruva could continue to produce about 25 to 35 kg of plutonium a year. This means that by 2010 the potential arsenal size could be about 130. Over the years, some 8,000 kilograms of reactor-grade plutonium may have been produced in the power reactors that are not under safeguards. Only about 8 kilograms of such plutonium are needed to make a simple nuclear weapon. If this spent fuel is declared to be for military purposes and not put under safeguards as part of the deal, India would have enough plutonium from this source alone for an arsenal of about 1,000 weapons, larger than that of all the NWS except the US and Russia.

In short, the nuclear deal not only promises to leave India's nuclear weapons capability intact but allows for a rapid and large expansion. It takes for granted that this is a good thing. The effects of the use of both the smaller yield fission weapons and the more destructive thermonuclear weapons in the arsenal are well known (Rajaraman, Mian and Nayyar 2004). Put simply, the smaller weapons will kill almost everyone within 1.5 kilometres of the explosion, the larger weapons will kill most people out to distances of 3.5 kilometres. The effects of radioactive fallout would spread many tens of kilometres further. Either kind of bomb would be enough to destroy a modern city.

The question that needs to be asked is: how many cities do Indian leaders wish to be able to destroy? There are many who believe that neither India nor any other country should have any nuclear weapons. These are instruments to create fear through the threat of genocide. The sixty years since Hiroshima should be enough to make clear to anyone that there is no security to be found in the threat to kill millions.

The US-India 'Global Partnership': The Impact on Non-proliferation[*]

Neil Joeck

The US and India have launched an ambitious new global partnership with strategic, economic, and energy dialogues. One component of the energy dialogue would allow the US to transfer nuclear technology to India, as the latter takes a number of non-proliferation steps, including measures to safeguard its civilian nuclear infrastructure. The civilian nuclear element of the new partnership requires that the US keep two balls in the air at the same time. Although the US wants to expand its bilateral relationship with India, it also wants to maintain its strong non-proliferation policy. Neither should come at the expense of the other.

In the eyes of many non-proliferation specialists, this new relationship rewards India for its recalcitrance regarding the NPT; it undercuts countries that accepted nuclear constraints; and it compromises longstanding US non-proliferation policy and the global non-proliferation regime. Such concerns are reasonable and deserve a thoughtful answer before implementing the new policy.

The history of non-proliferation policy has been one of adaptation and change. US policy goes back to the 1940s with the Baruch Plan and the Acheson-Lilienthal Plan. These early ideas for nuclear technology control met with resistance from the Soviet Union, so the US developed the Atoms for Peace approach. If the spread of nuclear technology could not be stopped, if bilateral measures were unavailable, then international monitoring might be a means of control. This approach did not stop new states from developing weapons, however, so the NPT was negotiated, incorporating some of the earlier approaches. India's nuclear test in 1974, shortly after the NPT entered into force, made clear that additional layers would have to be added to the regime. The Zangger Committee and the NSG were formed to restrict nuclear technology before it was transferred, rather than just monitoring its use after it was received. Congress added a number of elements to the non-proliferation regime by amending the Atomic Energy Act and the Foreign Assistance Act. The evasive actions

---

[*] The views expressed here are the author's own; they do not represent either the Lawrence Livermore National Laboratory or the U.S. Government.

of Iraq and North Korea made clear the need for the IAEA Additional Protocol. More recently, additional measures have been added such as UNSCR 1540, the Proliferation Security Initiative, and President Bush's enrichment and reprocessing proposals. Although the US must continue to implement policies that work, the history of non-proliferation policy shows that new contingencies frequently require policy adaptation or change.

It is necessary to look for new ways to achieve non-proliferation and security objectives: the agreement with India represents such an effort. The new policy does not require that the US abandon the NPT, the NSG, or any of the effective measures adopted over the years to stop proliferation. In marking its 35th anniversary, President Bush called the NPT the "key legal barrier" to nuclear weapons proliferation. The NPT remains a powerful multilateral security device that has enhanced international security. It has not eliminated all insecurities by any means—some states chose not to sign up, some that did have pursued nuclear weapons despite their commitments to the contrary, and the context for global disarmament remains elusive. So long as international insecurities and security competition persist, therefore, the world must find new ways to address them.

The new agreement with India recognizes that international security is achieved through a layered approach. The US has added to its non-proliferation and counterproliferation tool kit over the years. The agreement with India, while acknowledging the reality of India's nuclear weapons program, will supplement global efforts to enhance security. For years, India has been on the margins of the global non-proliferation regime. Indeed, India was a target of some of the measures cited above. Despite those efforts, Indian leaders concluded that they needed nuclear weapons to enhance India's security. Like other responsible powers, however, India has now committed itself to stopping proliferation by adopting many of the measures that the US values. The new agreement formalizes a cooperative relationship that will increase international security, thus addressing the fundamental goal of non-proliferation policy. The US has an opportunity to work with New Delhi on shared non-proliferation objectives as India takes steps to align its posture with prevailing international norms and practices.

The new relationship with India contains important advantages for international non-proliferation efforts. Looked at broadly, the US now has an additional ally in the international effort to restrict the flow of nuclear technology. One manifestation of India's new approach is its agreement to adhere to the NSG and MTCR guidelines. As India further develops its technology, ensuring that it takes part in international agreements to limit the spread of this technology will enhance international security. The agreement with India contains a second valuable element, in that it recognizes the value of safeguards and the role of the IAEA in ensuring against diversion of sensitive technology. India has accepted this norm by agreeing to separate its civilian and military facilities, agreeing to place safeguards on its civilian reactors, and accepting IAEA monitoring of the civilian facilities. A long-sought item on the international non-proliferation agenda has been to end fissile material production worldwide and to sign a FMCT. India's commitment to work with the US toward this longstanding objective represents another key advantage in the new partnership. Taken as a whole, these measures demonstrate India's endorsement of key non-proliferation objectives.

The price to the US for these changes and the inclusion of India as a member of the non-proliferation community (though not of the NPT) appears to be high. Congress must change or amend certain laws, which is no small accommodation. The 1978 NNPA, an amendment to the Atomic Energy Act, requires that a state adopt safeguards on its entire nuclear infrastructure before the US will transfer it any sensitive nuclear technology. It was adopted to achieve non-proliferation and national security objectives. With India having now agreed to place safeguards on its civilian program, the US must consider whether to change this law, thereby taking advantage of India's new thinking, or maintain the law and leave all of India's nuclear facilities unsafeguarded. If the law represents fundamental American values or principles, Congress should not change it. But changing or amending the law will not mean that the US or the rest of the world will incautiously transfer sensitive nuclear technology; it also will not mean that the US or other states will stop working to further global non-proliferation; it will not be the death knell for the NSG. Changing or amending the law would, however, provide an incentive for India not only to adopt objectives that the US values highly, but also become an active member of the non-proliferation community.

Having changed the Atomic Energy Act in 1978 to require full-scope safeguards as a condition of nuclear supply, the US pressured the NSG to adopt similar standards. The NSG finally did so in 1992; full-scope safeguards have been the standard for nuclear technology transfer ever since. A number of states that gave up nuclear ambitions are now members of the NSG and can be expected to demand to know why an exception should be made for India. The answer goes back to the goals of the NPT. Non-proliferation is at heart national security policy. Each nation that joined either the NPT or the NSG, did so as a sovereign state making careful judgments about how best to ensure its own and international security. Because of those decisions, the NPT continues to be the strongest and broadest multilateral security treaty in existence; the NSG continues to be a powerful tool for controlling the flow of sensitive technology; forgoing nuclear weapons continues to be the wisest policy choice for most states to enhance security. The new agreement with India does not alter those conclusions. Instead, the new agreement expands the list of countries committed to preventing further proliferation, thereby enhancing global security.

To conclude, US non-proliferation policy has changed over the years to meet new challenges to security. The new partnership with India provides an opportunity to increase global security while adapting to new conditions.

India's Rise to Nuclear Power Status and the Development of Canada's Nuclear Non-proliferation Policy

Ross Neil[*]

India's explosion of a so-called 'peaceful nuclear device' in 1974 made it the sixth country in the world to develop nuclear weapons and fixed its place in the history of global nuclear non-proliferation policy. Never far from any discussions of this history is the accepted fact that India utilized plutonium from a Canadian-supplied research reactor to fuel its first nuclear device. Less well understood, however, is that bilateral agreements limiting India's use of transferred nuclear technology and materials to peaceful purposes were as rigorous as any in place at the time.

Sober Second Thoughts – the 1974 and 1976 Policy Development Process

With the 1970 entry-into-force of the NPT and the development of a strengthened safeguards program at the IAEA which would cover all current and future nuclear activities in NNWS, Canadian officials remained concerned over India's decision not to ratify the NPT. In addition, India would only agree to the more limited facility-specific safeguards at its RAPS-1 and RAPS-2 facilities and would not agree to any safeguards at CIRUS.

In 1971, India conceded to a trilateral safeguards agreement with Canada and the IAEA that would see inspection duties transferred from Canada to the IAEA. Between 1971 and 1974, Canadian officials were apprised of India's diversion of spent fuel from CIRUS and intensified efforts to ensure that no diversion of spent fuel from the CANDU-type reactors would take place. But high-level talks failed to convince India that it should live up to the spirit of its over-arching bilateral agreement with Canada and sign on to the NPT as a NNWS.

Following India's 1974 test, Canada immediately suspended all nuclear cooperation with India and undertook an in-depth review of Canadian nuclear trade policy. The lesson Canada learned in 1974 was that the type

---

[*] The views expressed here are the author's own and do not represent official positions of the Government of Canada.

of assurances put in place between Canada and India, and the mechanisms for monitoring compliance with these assurances, were not adequate. The first phase of strengthened requirements came through additional controls placed on safeguarding exports of fissile material and non-nuclear materials (such as heavy water), as well as nuclear equipment and technology.

In addition to export controls, Canada removed any potential distinction between a 'peaceful nuclear explosive device' and nuclear weapons by requiring any recipients of Canadian nuclear materials and equipment to make a non-explosive use commitment. Conditions of supply would therefore be strengthened for any future bilateral nuclear cooperation agreements. Countries receiving nuclear goods from Canada would be required to seek prior consent for any re-transfer of Canadian-supplied nuclear items, for any reprocessing of fuel either supplied by Canada or used in Canadian-supplied facilities, or for storage and high enrichment of nuclear fuel. Conditions were also placed on the measures of physical protection and nuclear materials safeguards that would have to be in place for nuclear facilities and materials supplied.

Attempts were made by Canadian officials to renegotiate its nuclear cooperation agreements with both India and Pakistan to reflect the new strengthened requirements but the negotiations were unsuccessful. By May 1976, Canada had terminated all bilateral nuclear cooperation with Pakistan as well as India – suspensions still in force today. In December 1976, Canada further strengthened its non-proliferation policy by restricting any new nuclear cooperation to those NNWS that had ratified the NPT or made an equivalent binding commitment, and thus had accepted IAEA safeguards on the full-scope of any nuclear fuel-cycle activities taking place.

By 1978, the NSG had formed as a group independent of the NPT and had published its own list of export control guidelines. The Zangger and NSG lists provided further impetus for Canadian policymakers and nuclear experts to undertake their own in-depth examination of all components and materials required for the CANDU reactor system.

Balancing Proliferation with Nuclear Safety

Canada's implementation of its 1974/1976 policies and strict adherence to multilateral and domestic export controls continued unabated throughout the 1980's, despite emerging indications that safety was being marginalized at the Canadian-supplied CANDU facilities in India. Serious pressure tube failures experienced at both the Pickering and Bruce nuclear generating stations in Canada raised concerns that similar problems could occur overseas. Hard-line export controls conflicted with the views of scientists and engineers who cautioned the Canadian government not to compromise reactor safety for political ends. The 1974/76 policies left open no windows for supplying spare parts or even for association of Canadian nuclear experts with their former counterparts in either non-NPT country. But Ministerial reviews consistently maintained no distinction between safety assistance and other forms of nuclear cooperation in Canada's nuclear non-proliferation policy.

While trust-worthy data is difficult to obtain for the early years of reactor operation in India and Pakistan, there is little doubt that political decisions to isolate these countries resulted in dangerous operating conditions at nuclear facilities there, more than likely contributing to dangerously high radiation doses for workers and the public. By 1988, no cooperation whatsoever had taken place between Canada and India (or Pakistan), despite repeated attempts by nuclear industry officials and nuclear safety engineers to have a one-time exception built into the policy to address safety issues, or to revamp pre-1976 agreements with India and Pakistan to allow some form of safety-related collaboration.

Critics of the Canadian policy focused on two key arguments. The first was that ostracizing India and Pakistan ignored the serious nuclear safety hazards that might emerge at the Canadian-supplied reactors after cooperation was withdrawn, which were not anticipated due to the assumed robustness of the CANDU design. The second key argument was that isolation actually enabled India (and to a lesser extent Pakistan) to become self-sufficient in its nuclear fuel-cycle activities, raising its prestige beyond 'nuclear weapons capable state' to that of a country that could perhaps even teach Canadian experts a thing or two about how to maintain an aging reactor without the best available technology.

Toward Nuclear Collaboration, but not Cooperation

Commencing in 1989 and continuing into 1990, Canada's nuclear non-proliferation policy began adapting to the moral imperative that Canada's nuclear safety specialists felt they had to ensure the safe operation of the Canadian-supplied nuclear facilities. The challenge facing Canadian policy-makers was to balance this imperative with Canada's international obligations through its NPT and export control group memberships. Toward this end, Canadian nuclear scientists were given approval to participate in a series of safety missions to inspect and report on the safety issues at Canadian-supplied facilities in India and Pakistan. A path forward was found in the distinction made between *cooperation* and *collaboration*, where indirect Canadian assistance could be provided on a case-by-case basis through multilateral auspices such as WANO, or the CANDU Owners' Group (COG, an industry association comprised of Canadian and foreign operators of CANDU technology), or the IAEA's technical advisory committees.

As the international inspections revealed serious concerns, Canada agreed, within the parameters of its existing nuclear non-proliferation policy, to permit Indian and Pakistani participation in the information-sharing program of the COG. The program would allow both countries access to public domain, non-proprietary information relevant to the safe operation of Canadian-supplied (and IAEA safeguarded) CANDU reactors. While India and Pakistan had arguably always had access to public domain information, COG offered a means to bring India and Pakistan into a formal group where lessons and best-practices could be shared between members.

In May of 1990, the Government of Canada agreed to offer a limited program of nuclear safety assistance to the three Canadian-supplied CANDU reactors in India and Pakistan that were under facility-specific IAEA safeguards. Support would be strictly limited to urgent and critical safety-related problems for which Canadian-specific expertise was required. Projects undertaken would be reviewed, approved and monitored by Canadian government officials, but would be undertaken by Canadian experts under multilateral auspices (i.e. the IAEA) with no direct bilateral assistance or Canadian funding. India subsequently requested that the program be made available for both its safeguarded and un-safeguarded reactors. Canada was unwilling to accept this condition, leading to India's refusal of the offer in 1992.

The 1998 detonation of several nuclear explosive devices by both India and Pakistan made the continued quest for a dynamic policy balance between nuclear non-proliferation and safety all the more challenging, curtailing any notion that Canada-India bilateral nuclear discussions could take place in the immediate future. But in recent years, industry groups have increased pressure on the Canadian government to allow insights to be gained from India's now significant CANDU experience and to seize what is perceived as a lucrative market for nuclear energy in South Asia. Under Minister of Natural Resources Herb Dhaliwal, who had deep family connections to India and also served as the Minister responsible for AECL, several opportunities emerged for a thorough re-thinking of the

1974/76 policy with its 1989/90 expansions. However, the other two Canadian ministers responsible for implementing nuclear non-proliferation (Foreign Affairs and International Trade) rebuffed proposals to have the nuclear industry included in official Canadian trade missions to India. The dire need by India for nuclear technology, expressed by Canadian industry as a significant market opportunity, was viewed by non-proliferation officials as a sign that the international community's unified approach to isolate India had been working and should continue.

Under a subtle Canadian policy expansion in 2001, Indian and Pakistani officials are now permitted to participate in safety expert review missions at nuclear facilities in Canada, a bold proposal that survived heightened concern over security at Canadian nuclear facilities following the attacks on the US in September 2001. Review missions have been authorized under the proviso that the visits are multilateral and safety-related. An additional expansion allows for bilateral consultations between Canadian and Indian government officials, a significant departure from the hard-line isolationist stance taken by Canada since 1974. Such talks, if pursued, would provide a forum for Canada to remind India of its standing offer of limited nuclear safety assistance. Direct regulator-to-regulator contacts would no longer be discouraged under this new approach, which could contribute to the improvement of the safety culture in India.

Conclusion

Uncompromising positions in the UN Conference on Disarmament have been responsible for much stalled progress toward the negotiation of a FMCT. The status of both India and Pakistan (along with Israel) as *de-facto* NWS that have not acceded to the NPT poses a direct challenge to the universal implementation of the treaty which, despite its imperfections, remains a cornerstone of global disarmament and international security, and the central instrument with which Canada works to achieve the objectives of both nuclear non-proliferation and disarmament. Canada's status as a major global player in the nuclear supply chain and its special – though troubled – nuclear relationship with India provide a valuable entry point for formal bilateral discussions on energy matters, including issues of nuclear trade. First, however, clearly articulated objectives must be communicated that do not undermine historical efforts to improve global security through non-proliferation. With significant Indo- and Pakistani-Canadian populations, and a tolerant, inclusive, multicultural society, Canada should give high priority to re-establishing friendly relations between and with the sub-continental neighbours.

The Impact of the US-India Joint Statement for Canadian and International Nuclear Non-proliferation Efforts

Ronald E. Stansfield[*]

The US-India Joint Statement of 18 July 2005 marks a pivotal point in the history of our efforts to control the proliferation of nuclear weapons, and to move toward their ultimate elimination. The repercussions of the Joint Statement are going to be felt for years to come. The question of whether these repercussions will be positive or negative will depend ultimately upon the manner in which we – Canada, our allies and partners, the entire international community – respond to this major shift in the nuclear non-proliferation landscape.

Canada-India Nuclear History

Canada's nuclear relationship with India began almost exactly 50 years ago. In September 1955, Canada signed a nuclear cooperation agreement with India. The immediate impact of the agreement was the supply of the 40 MW CIRUS research reactor in 1956. This was followed by a further agreement in 1963 to construct a small 200 MW commercial nuclear power plant in Rajasthan, which entered operation in 1972 (RAPS-1). In 1966, Canada agreed to provide a second nuclear power plant (RAPS-2), the construction of which had not been completed when Canada suspended nuclear cooperation with India following the May 1974 nuclear test.

Both RAPS 1 and 2 have been under IAEA safeguards since their initial operation and have therefore not contributed to India's nuclear weapons programme. Unfortunately, however, India, without Canada's agreement and without acknowledgement or compensation, has replicated the CANDU design and built a further 11 plants of this type, none of which are under IAEA safeguards.

---

[*] The views expressed here are the author's own; they do not represent either the Department of Foreign Affairs and International Trade or the Canadian Government.

The May 18, 1974, Indian nuclear test was extremely traumatic for the Canadian Government. Nuclear cooperation with India was suspended immediately and then formally terminated in 1976. It has not been renewed for over thirty years. More importantly, the Indian experience formed the fundamental basis and rationale for Canada's current nuclear non-proliferation policy. This is the irony of the conundrum currently facing us with regard to the US initiative to now renew nuclear cooperation with India. The very country that was the source of Canada's policy is now threatening its continued viability.

Before Canada will consider entering into nuclear cooperation with any NNWS, that state must:

- make a legally binding commitment to nuclear non-proliferation by becoming a Party to the NPT, or an equivalent, internationally legally binding agreement; and

- thereby accept full-scope safeguards by the IAEA on all of its current and future nuclear activities.

In addition, any countries wishing to enter into nuclear cooperation must conclude a legally-binding bilateral Nuclear Cooperation Agreement with Canada which includes, among other things, certain additional nuclear non-proliferation commitments.

The requirements of Canadian nuclear non-proliferation policy, and in particular the commitments set out in the bilateral Nuclear Cooperation Agreements, apply to all items exported directly from Canada or indirectly via third parties. They also apply to non-Canadian equipment or nuclear material used in conjunction with or produced from supplied Canadian nuclear items, and to equipment manufactured on the basis of technology provided by Canada or through reverse engineering.

Therefore, under current Canadian policy, substantive renewed nuclear cooperation with India, in terms of trade in major nuclear items, is not possible given India's official status. Moreover, the NSG adopted Canada's standard in 1992, and therefore all 45 members of the NSG apply this standard. The NSG, in fact, like Canada's nuclear non-proliferation policy, was formed in 1974 as a direct reaction to India's first nuclear test - another irony given the stated US intention that it will seek an exemption from the NSG Guidelines for India.

Behind the US-India Deal

As important as it is to note what the July 2005 US-India Joint Statement contains, it is equally important to note what it doesn't contain. Careful examination of the language of the document reveals a number of significant qualifiers and ambiguities that will make renewed cooperation a difficult and lengthy process. First, the US has not agreed to renew nuclear cooperation with India, but only to "work toward" this goal. To be able to move forward, the US has to change at least four major pieces of national legislation, including the 1979 Nuclear Non-proliferation Act. It has to negotiate a bilateral nuclear cooperation agreement with India, which must be ratified by Congress. It has to either change the NSG Guidelines or seek a special exemption for India. These are all important obstacles.

The reference by India to reciprocity in the Joint Statement indicates it would expect such from the US. The US has, however, repeatedly said since the Joint Statement that it is not prepared to recognize India as a NWS. The notion of reciprocity also begs the question of who goes first. While US officials have indicated that India is expected to take the first steps, by contrast, public statements by Indian officials indicate that India clearly expects the US to move first.

With regard to the separation of its military and civilian nuclear fuel cycles, India has not indicated what facilities would be included in each subset, or how this separation will unfold. The reference to India "voluntarily" accepting IAEA safeguards also implies that it expects to be treated equally to the five recognized NWS in this regard. India's commitment to maintain its unilateral moratorium on testing is a step back from previous statements indicating that it would eventually sign the CTBT. It is disappointing that in the Joint Statement India did not declare, as the five established weapon states have done, a unilateral moratorium on the further production of fissile material for weapon purposes. India has still not formally adhered to either the NSG or MTCR Guidelines, despite the commitment in the Joint Statement to do so.

The greatest impediment may ultimately be the question of India's status. However, the logic of the US approach to India only holds together if it accepts that India is indeed a NWS. How can one, for example, assert India is a NNWS while recognizing that it has civilian and military nuclear fuel cycles? How can one make an exemption from the NSG Guidelines for one so-called NNWS while insisting that the Guidelines continue to apply to all other NNWS?

Canada-India Relations: The Future

Canada's broader bilateral relationship with India has been evolving in recent years. The Joint Declaration by Canada and India issued during the Prime Minister's visit to India in January 2005 underlined the significance both countries attached to initiatives that strengthened the Canada-India partnership and contributed to addressing global challenges more effectively. Canada's *International Policy Statement* tabled in Parliament in April 2005, specifically identified India, among others, as an emerging economic power which would be one of the key drivers of a new era of global economic growth. Canada, as a consequence, would increase the pace of its economic engagement with that country.

Based on recent further positive actions by India, including its support for the September IAEA resolution on Iran's nuclear activities, the Government believes that circumstances have evolved sufficiently to allow Canada to re-engage India on a measured, reciprocal basis in the nuclear field. At the same time, it should be clearly understood that the Canadian initiative announced on 26 September 2005 does not constitute a substantive change in existing Canadian nuclear non-proliferation policy. The initiative does seek to underline the importance Canada places on India as a partner, and to respond to India's stated commitment to central nuclear non-proliferation norms. While we welcome the recent commitments India has made to nuclear non-proliferation, we expect them to do more, and will urge them to move further into the nuclear non-proliferation mainstream. In the interim, the current policy of precluding transfers of especially designed or prepared nuclear items subject to the requirements of Canadian nuclear non-proliferation policy will therefore be maintained.

While it remains to be seen if the US will be ultimately successful in renewing full nuclear cooperation with India, the situation is forcing a re-examination of how we perceive the nuclear non-proliferation regime, and how best to achieve its objectives of preventing the further proliferation of nuclear weapons and moving toward their eventual elimination. US Under-Secretary of State for Political Affairs Nicholas Burns recently indicated to the House International Relations Committee that the US had "surrendered to reality" regarding its relationship with India. The world may be forced to move away from the traditional NPT-centric approach to non-proliferation, which may not necessarily be a bad thing, as long as we carefully prepare the necessary steps to manage the consequences.

This approach also applies to Canadian national policy. We have signalled to the Indians our willingness to re-establish the relationship without making any new commitments that would compromise that policy or the principles behind the policy. The ball is, however, clearly in India's court. The nature of the commitments they demonstrate to nuclear non-proliferation, arms control and disarmament and to international security issues generally in the period ahead will determine whether or not we move forward on the nuclear file.

Mixed Consequences of the Indo-US Nuclear Deal

S. Paul Kapur

Although some observers have come out strongly for or against the proposed Indo-US nuclear cooperation agreement, I believe that its impact is likely to be mixed; the proposed deal will have a range of costs and benefits between different issue areas. I discuss the deal's advantages and disadvantages in three spheres: economics, diplomacy and security, and non-proliferation. It offers modest, though potentially important advantages in the economic realm; creates a number of short-term benefits and longer-term risks in diplomacy and security; and could inflict significant damage on the non-proliferation regime.

Let us look first at the economic implications of the deal. India is already a significant economic power; its economy is the world's sixth largest, with a GDP of over $3 trillion (in US dollars, measured in purchasing power parity) (Central Intelligence Agency 2004). Its economy is growing at about 8% annually (G. Srinivasan 2005a, Wilson and Purushothaman 2003). This growth is good for other states; a prosperous, productive India offers an enormous market, and will provide goods and services to international consumers at competitive prices.

A by-product of this expansion will be a dramatic increase in energy consumption. In 2002-2003 India produced 639 terawatt hours of electricity. Current projections indicate that by the middle of this century, India will consume approximately 8,000 terawatt hours of electricity per year (M. R. Srinivasan 2005b). Although at present India meets energy needs overwhelmingly through fossil fuels, doing so is expensive and contributes to greenhouse emissions. Nuclear energy offers a partial solution; it could cleanly and inexpensively provide roughly a quarter of India's electricity by the middle of this century (M. R. Srinivasan 2005b).

Are there any economic downsides to US-India nuclear cooperation? India's civilian nuclear energy program may not live up to its potential. The program certainly has had a disappointing past. The Atomic Energy Commission announced in 1954 that nuclear power would generate 8,000 MW of electricity by 1980-1981. However, by 1970, nuclear plants were generating a mere 420 MW of electricity. At this point the Commission revised its projections downward, claming that by 1980-1981 Indian nuclear plants would generate 2,700 MW of electricity – roughly the amount of electricity that power plants are producing today (Perkovich 2005). Even if nuclear power lives up to optimistic projections, its impact will be limited, as the majority of Indian electricity production will be coal-driven. And nuclear energy will not be able to power cars, buses, and trucks, which are India's primary petroleum consumers (Carter 2005). A second problem is that nuclear power creates a small possibility of a catastrophic accident or terrorist event, outweighing the environmental benefits of reduced greenhouse gases.

These problems are significant, but do not wholly negate the economic benefits. Despite failures in the past, access to new technologies and an ample fuel supply should make nuclear power production more efficient. And the risk of a major accident or attack, though real, is small; India has not experienced such an event in the past. New nuclear equipment and technology, which will begin to replace the old Indian infrastructure, will reduce the risk of such an occurrence even further (Ganguly 2005).

Let us now turn to the diplomatic and security arena. Here the effects of the Indo-US nuclear deal are more mixed. American cooperation despite India's development of a nuclear weapons capacity – cooperation that will require change in domestic laws and alteration of international non-proliferation rules (McGoldrick, Bengelsdorf and Scheinman 2005)– marks a clear end to the period of "nuclear apartheid," when the US sought to prevent India from acquiring nuclear weapons (Singh 1998). Indeed, the US is now treating India as a *de facto* NWS. This should boost Indo-US relations in the short term. The longer-term danger is that US leaders may behave as if the nuclear deal has bought them India's allegiance. Some American policymakers clearly believe that in return for the agreement, India is obliged to support US non-proliferation efforts toward Iran. As Congressman Tom Lantos put it, "There is *quid pro quo* in international relations. And if our Indian friends are interested in receiving all of the benefits of US support we have every right to expect that India will reciprocate in taking into account our concerns" (Anon 2005b). Many Indians resent this view (Karat 2005). A similar situation could emerge regarding US policy toward China. The Bush Administration hopes that greater Indian economic and military prowess will be helpful in the containment of Chinese power. However, US and Indian objectives vis-à-vis China may not always be compatible (Perkovich 2005).

On the security front, the deal may in the near term augment Indian military capabilities. First, despite speculation that the agreement's separation of civilian and military nuclear programs might effectively cap the size of the Indian nuclear arsenal, it could actually have the opposite effect. India must currently use its uranium to fuel both its civilian and military programs. The proposed deal will grant India access to international fuel supplies for its reactors, enabling it to devote its indigenous uranium supply to its military program. Significantly, a larger Indian nuclear arsenal may not simply be an unintended consequence of the agreement – it may be part of a broader Indo-US strategy to afford India a more robust deterrent against China (Mian and Ramana 2005, Perkovich 2005). Second, the proposed nuclear agreement could encourage more conventional Indo-US military cooperation, giving India access to cutting-edge systems and weaponry, even though such conventional arrangements are not explicitly part of the nuclear deal. Although these factors should boost India's defence capabilities in the short term, their longer-term effects are as yet unclear. Increased Indian military capabilities could lead to Chinese and Pakistani conventional and nuclear arms racing, and greater Sino-Pakistani cooperation.

Finally, what impact will the proposed agreement have on global non-proliferation? It seems clear that the deal will damage the NPT. The NPT offers civilian nuclear energy only to states that agree to forego the development of nuclear weapons. India, however, remains outside of the treaty, and yet acquired a nuclear weapons capacity. For the international community now to share civilian nuclear energy with India weakens the NPT's material inducements. Indeed, the proposed agreement could encourage current NPT signatories to opt out of the treaty and embark on a weapons program. Additionally, it could undermine efforts to prevent other nuclear supplier countries from providing nuclear technologies to potential proliferators such as Iran.

The proposed agreement also implicitly rejects the NPT's larger worldview – namely that all nuclear proliferation is equally undesirable, and that the roster of nuclear and non-nuclear states should remain frozen as it was in the early 1970s. Underlying the deal is the view that a state as economically and militarily powerful as today's India, situated in as volatile a region as South Asia, cannot be expected to forego nuclear weapons forever. Moreover, given the Indian record of responsible nuclear stewardship, sharing civilian nuclear technology is unlikely to give rise to future proliferation dangers. Thus, according to the logic of the proposed agreement, not all newly nuclear states are created equal, and they should not be treated in the same manner (Anon 2005e, Perkovich 2005).

Even as it undermines aspects of the NPT, the agreement seeks to bring India into the larger non-proliferation regime. It requires India to separate its civilian and military nuclear programs, and to place its civilian program under IAEA safeguards, including the Additional Protocol. Although military facilities would remain off limits, this could be a major improvement over the status quo, since at present no aspect of the Indian nuclear program is open to international scrutiny. The deal also would commit India to secure nuclear materials and technologies and prevent their spread, continue its moratorium on nuclear testing, and to participate in the negotiations of a FMCT.

This array of non-proliferation measures is impressive at first glance. However, close examination reveals that non-proliferation benefits are less substantial than they would seem. The deal's safeguard component suffers from two important problems. First, the division of military and civilian nuclear programs is up to India; no independent rules exist to determine how facilities will be classified. Second, the agreement allows India to decide what type of safeguards will be applied to civilian facilities. If India chooses to adopt voluntary safeguards, rather than safeguards in perpetuity, it will be able to withdraw any facility for national security reasons, moving a previously civilian reactor into the military sector (though such voluntary safeguards could prevent nuclear cooperation with the US unless Congress changes current law) (McGoldrick, Bengelsdorf and Scheinman 2005).

The proposed agreement's other non-proliferation components are important, but they require India to take steps that it was previously taking anyway. India was already required to prevent the spread of nuclear materials and technologies by UN Security Council Resolution 1540. And India was already observing a unilateral moratorium on testing, and supporting the FMCT. Significantly, the proposed agreement does not require the Indians to halt fissile material production – a step that the other five official NWS have voluntarily taken. Thus, between now and accession to the FMCT, India can continue to produce fissile material unhindered, creating pressures for similar behaviour in other countries and increasing the dangers of nuclear theft or accident.

The likely effects of the proposed Indo-US nuclear agreement are thus mixed. The deal should result in important if modest international economic gains, a combination of short and long-term benefits and dangers in the diplomatic and security realms, and harm to the non-proliferation regime. It is difficult to argue that a state of India's size, economic and military trajectory, and strategic location, should be denied nuclear weapons. And yet an acknowledgement of the reasonableness of India's claim to a nuclear weapons capacity is necessarily in tension with efforts to prevent the spread of nuclear weapons. Even if we accept that the nuclear status quo was not viable, it seems that the deal could have been more ambitious on the non-proliferation front.

The India-US Nuclear Accord in Strategic Context

T. V. Paul

I will focus on the larger political issues surrounding the India-US nuclear accord of July 2005. In sum, I see the accord as having mixed implications for the non-proliferation regime and for India's integration into the world nuclear order.

As for the regime, it does generate some questions about equity, fairness, and the treatment of different new nuclear states and non-nuclear states by the US and other nuclear haves, and their commitment to non-proliferation principles and norms. The question naturally arises: is India so special or unique that the Western countries are willing to accommodate it while they are fiercely opposed to Iran and North Korea acquiring nuclear weapons? This is further compounded by the fact that the US is fighting a war in Iraq, the stated reason being the alleged nuclear acquisition efforts by Saddam Hussein.

I would argue that the likely impact of the US-India nuclear accord on the non-proliferation regime is minimal and unlikely to lead to additional states acquiring nuclear weapons. This is partly because India is not a member of the NPT and is unlikely to join it as a non-nuclear state. Most regional states acquire or give up nuclear weapons because of situational factors: i.e. due to regional and domestic politics reasons, and in recent years due to their fear of US intervention. India's accommodation with the regime may have little impact even though some countries might use it to justify their position.

India is different. India's nuclear acquisition occurred over a period of 30 years. India was a reluctant proliferator. It chose the path after struggling with the unequal nuclear order that was thrust upon it by the great powers that could not allay India's security concerns arising out of China's nuclear acquisition and China-Pakistan nuclear collaboration.

India is essentially a status quo power; defensive and reactive. It has been at times critical of the international nuclear order, but has restrained itself from spreading the weapons or materials to other states. Its nuclear doctrine is based on *no first use* and it has kept the components of nuclear weapons separated -- ensuring that the weapons are not fired haphazardly. Unfortunately, Pakistan has a first use doctrine and has assigned nuclear weapons a number of roles -- six to be precise -- that include use if there is an economic strangulation by India. Since its nuclear acquisition, Pakistan has also intensified its asymmetric war in Kashmir, hoping to upset the territorial status quo under the cover of nuclear protection.

This generates a bigger issue. When it comes to nuclear proliferation, there is a difference between a status quo power and a revisionist state. The non-proliferation regime is agnostic on this political dimension. Why does the world shudder at nuclear acquisition by Iran and North Korea? Because they are middle-sized revisionist states -- they have a propensity to upset regional orders and they have active conflict relationships with the great power system and with their neighbours. In many respects, Pakistan is also part of this revisionist group of states. It no longer makes sense to hyphenate India and Pakistan in this regard as a pair.

Pakistan pursues both territorial and ideological/religious revisionism. And nuclear weapons and active territorial revisionism form a deadly combination. Due to the theocratic nature of Pakistani society, it has been quite unhappy with the non-Islamic world, although its elite may form strategic alignments with the West. Sections of its elite are also willing to sell nuclear materials for commercial and ideological purposes. The military elite of Pakistan follow a highly ambitious realpolitik approach toward national goals and problem solving. Nuclear possession has increased their willingness to take high risks and engage in asymmetric strategies.

Another big problem with the NPT is that it assumes power transitions in the international system will not take place and that the P-5 will eternally remain as the top actors. The regime is built around sovereign inequality -- or more precisely two types of sovereign equalities: one among nuclear haves and the other between nuclear have-nots. It has no room for the orderly exit of a declining power or the entry of a rising power. One of the reasons for the NPT's adoption was that the nuclear haves promised nuclear disarmament as part of the grand bargain in 1968. But with the extension of the NPT in perpetuity in 1995, the P-5 has little incentive to link vertical and horizontal proliferation.

Status inconsistency could emerge as a major source of tension in the international system due to differential growths of countries. The NPT could contribute to this problem if it cannot find a way to accommodate rising powers. The treaty has survived so far partly because no major power is outside it.

India is perhaps the only candidate in the near term for a major power position. It just happened to be a latecomer. In every indicator of hard and soft power, India has no parallels. Brazil comes close on some measures but with a population of 151 million and in the strategic backwaters of South America, its chances of gaining a leadership role are not too high.

The Indian elite seems to have concluded (rightly or wrongly) that major powers such as US and China will not take India seriously until it achieves a certain capability threshold. Even when it opposes India's nuclear ambitions openly, China's contempt for India has been based on an expectation that India will never catch up economically or militarily with China. There is a domestic consensus in India that in order to be taken seriously India has to achieve both economic and military power, although in the medium term, substantial distributional problems and social maladies will act as a drag on India's ambitions.

India's domestic political situation is quite unique, partly because of its unwieldy democratic system. Whenever the non-proliferation regime managers imposed new restrictions, the Indian reaction has been defiance with an increase in nuclear and missile activity. Past efforts to cap the Indian program through the NPT or CTBT aroused intense nationalism in India. This form of nationalism is much stronger in India than in any other near-nuclear countries and it is very much tied to Indians' notions of national independence and their peculiar colonial history. No Indian political party will ever agree to India's unilateral nuclear disarmament unless it is part of global disarmament. That is one reason why India is quite different from other NNWS.

If the non-proliferation regime has not succeeded in forcing India to accede to the NPT so far, what guarantee is there that it will do so in the future? Would it not make sense to limit the Indian nuclear program -- by separating the civilian and military components and thereby restricting the number of weapons that India can build when full disarmament may not happen?

It is better for India as well to have some non-proliferation norms rather than none. India needs the regime if it wants to become a *status quo* state and integrate itself into the international system. But the regime needs to adapt to changing circumstances. Keeping India outside of it, while maintaining normal nuclear trade with China (the US, Canada and the EU are all selling civilian reactors to China) -- even when China has blatantly violated its NPT obligations by transferring nuclear materials to Pakistan -- generates a bit of a credibility problem. It is better to manage India's rise peacefully and integrate it into the nuclear order than keep it as an outsider. Making India a stakeholder in the regime and a supporter of its enforcement for regional states would strengthen global peace and order. Smaller states will eventually adjust as they did with China. Democratic India's integration as a lead actor is good for all principal states concerned -- it is bound to become a major economic and military force in the next 10-20 years, if the current growth rate continues.

Nuclear reactors are very much part of India's energy plans. Fourteen reactors are now in operation in India, nine are under construction, and four more are planned. Only six are currently safeguarded. It would be better if India built safe nuclear reactors that were safeguarded against diversion. It would also be better to use nuclear energy so that India does not need to burn more coal and upset the global ecological system and put more pressure on world-wide oil and gas prices. Further, if India's plans for thorium-based reactors succeed, they will be under no safeguards. Under the agreement with the US, India will have to place all civilian reactors under international safeguards.

Relations between countries like Canada and India will not improve to their fullest potential without a rapprochement on the nuclear issue. This issue has plagued economic relations as well for over thirty years. How long do we want to hold hostage to a single issue the tremendous potential for economic and political relations? Similarly, in US-India relations, the nuclear issue has perhaps been the single most important issue for the past thirty years or so that prevented full rapprochement of the two estranged democracies.

One drawback of the US-India agreement is that it does not mention nuclear safety and waste dumping. Although India's safety record is fine so far, it is a matter of concern, partly because India suffers a high level of terrorism. We do not know how secure India's civilian and military and command and control facilities are against terrorist attacks. Another issue is waste dumping. What India is doing in this regard is not clear.

India's recent vote at the IAEA on Iran shows India's ambivalence with respect to another nuclear power in the neighbourhood. But India's non-proliferation philosophy has to evolve a bit more creatively. In principle, Iranian nuclear capability is against India's interests, partly because India should not want a revisionist neighbour. India may have to join ranks with the West so as to dissuade countries like Iran and North Korea, given all the challenges these states create for their regions and the world. India has not made much effort to work as a go-between to find a solution to the problem.

I would conclude that bringing in India as a stakeholder of the regime is in the longer term beneficial for the regime itself and for the cause of non-proliferation, although it does create short-term legitimacy issues.

Discussion Summary

Discussion Summary

(Editors' note: Following is a comprehensive summary of the main points raised during the discussion period of the conference. It is not a verbatim transcript. Conference panellists have been given the opportunity to review this summary to ensure accurate rendition of their comments.)

The discussion in the afternoon session commenced with the Chair, Wade Huntley, presenting some questions emerging from the morning's presentations:

- Advocacy of nuclear cooperation with India is premised on the idea that India is a "responsible" nuclear weapons power. What does this mean? Does discrimination among states by such criteria help or hinder non-proliferation?

- Will the US-India deal set precedents relevant in other contexts? What will be the extent of this precedent-setting function? And will such precedents have positive or negative effects? For example, does North Korea get the message, "To gain stature, keep nuclear weapons and hold out" or "To be embraced by the global community, act responsibly?"

- How meaningful is the planned safeguarding of Indian nuclear facilities? Will such provisions in the deal change Indian behaviour?

- How much did concerns over the rise of China influence the deal? How big a role should this factor have?

- What can we predict about the future role of the NPT in broader non-proliferation efforts? Is the NPT still a non-proliferation centerpiece or an anachronism?

- Will renewed international support for India's civilian nuclear sector strengthen or weaken NSG guidelines? Will this agreement have the same implications for the NSG as for the NPT?

- What is Canada's policy? Given uncertainties in the policies of the US and other countries, what trajectory should Canadian policy take?

- Can the NPT and the non-proliferation regime it represents adjust to a changing global order? Or is the regime too inflexible to accommodate the ebb and flow of global power and influence?

Seema Gahlaut began the discussion by observing that the NSG was actually not intended to target India as is commonly thought. Proof of this can be found in Kissinger's memoirs. In fact India had even been asked to join the NSG at an early stage. It was recognized that India's program would proceed apace regardless of sanctions and the group's aim was to prevent other countries from testing. As of July 2005 India has incorporated the NSG and MTCR restrictions into its domestic legislation.

Ron Stansfield agreed that the NSG had not been set up to punish India, although it was spurred by India's 1974 nuclear test. He also agreed that it makes sense to invite India to join the NSG, not merely exempt it from NSG rules. While India has changed its domestic legislation, the country has not officially notified the IAEA of its adherence to guidelines. France, Brazil and Argentina were not NPT members when they joined the NSG. Possibly India could also be accommodated. The Zangger Committee (which is part of the NPT) could also be expanded. In any event, the NSG must be reformed. Stansfield also suggested that instead of bending the rules to accommodate India, the world should take up the suggestion of the Director-General of the IAEA and create a convention open to all nuclear exporters. This would eliminate the problem of the inclusion of non-NPT parties who exported nuclear technology.

Ernie Regehr urged the participants to clarify what was implied by making India a "stakeholder" in the non-proliferation regime. Is the imposition of safeguards enough to bring India into accord with the regime? Gahlaut replied that the agreement would transform India's unilateral undertakings into bilateral agreements. She added that IAEA safeguards were not intended to stop the building of weapons, only to ensure that material intended for peaceful cooperation was not diverted to military uses. The deal would ensure that at least some of India's existing reactors would be brought under safeguards.

Regehr predicted that the US would be unable to get the Indians to agree to restrictions on fissile material production. He feared that the price for India's entry into the order was being paid by the NNWS that had renounced their nuclear capabilities. The payoff for the deal went to the US, which was creating an Indian surrogate for itself in Asia. What's in it for the NNWS?

Stansfield agreed with Regehr that safeguards on Indian facilities would not serve the cause of non-proliferation if total fissile material production was not capped – ideally through the FMCT, and especially if Indian domestic uranium production is not covered. If the IAEA safeguards were to be enforced on the 70% of the program devoted to civilian purposes without impinging on the rest of India's nuclear endeavours, the deal would merely free up plutonium to be used in weaponisation. Under these circumstances, it would be a waste of the international body's resources.

Gahlaut agreed that the Indian government would now have a greater quantity of plutonium available for manufacturing nuclear bombs. The intentions of India's DAE are unknown. However, she argued, a sharp increase in the arsenal was unlikely because Indian decision-makers were apt to think of nuclear weapons as political instruments of power rather than military weapons. Larger issues are more important. Safeguards are intended not only to stop new nuclear states from emerging, but also to limit and create accountability among existing nuclear states. Safeguards in India will ensure that all nuclear material imports will be used peacefully.

On the issue of the NPT and the health of the regime, Stansfield observed that the US actions were aimed at tackling the institutional inertia of the NPT. Just as it had repudiated the ABM treaty on the grounds that it was outdated, the US saw the NPT as a treaty of its time and no longer very useful. In the US view, there is a need for a new regime with new norms. There are now 183 countries locked out of nuclear weapons possession – the key is reducing nuclear weapons arsenals in the other nine.

T.V. Paul noted that the NPT is not an apolitical treaty. Rather, it is the product of bargaining and great power imposition. In the US in particular, strategic interest often overrides arms control goals.

M.V. Ramana, along similar lines, suggested that the US is the revisionist state when it comes to the NPT. The NPT is a means to the end of disarmament – the US instead wants the NPT to be only an instrument for non-proliferation. The deal reinforces the centrality of the nuclear issue to India's foreign policy, but testifies to the influence of India's nuclear lobby rather than to the salience of nuclear weapons in domestic politics. The NPT is indeed discriminatory but the answer to that is not to add to the discrimination. This deal with India impinges on the NPT's principal goal of disarmament; therefore it is bad for the NPT.

Gahlaut contested this point and said that the NPT was *not* a disarmament treaty. In fact the only real disarmament treaty is the Chemical Weapons Convention, which provides for the destruction of existing stockpiles.

Neil Joeck described both the NSG and NPT as attempts to limit the transfer of nuclear technology. If their goal is to reward states who support non-proliferation norms India is a great candidate. If the goal is to encourage other states to adopt such norms, the deal is a step in that direction. The deal was necessary because the NSG was turning into a device for punishing India, which is a legitimate but isolated member of the international non-proliferation community. It is pointless to continue to ostracize a country for its actions thirty years ago. The point at that time was to try to keep India from retaining its nuclear weapons; that didn't work. The deal is not just about non-proliferation; there are other, broader elements.

Joeck also pointed out that at the NPT review conference the US had tried to have a discussion on violations of the treaty by North Korea and Iran, but found no other countries interested, and was frustrated at not getting these items on to the agenda. He observed that the NPT is a limited document that is too hard to amend, so we have to think broadly. He also noted that the US nuclear arsenal has shrunk dramatically thanks to arms reductions.

Ross Neil observed that India had not broken any multilateral treaties. The only bilateral agreements it may have broken with its 1974 test were those with Canada and the US. He also noted that all treaties, like clubs, discriminate or have conditions or costs for membership; but states are free to stay out of them.

On the role of China, T.V. Paul observed that a major American motivation for the deal was getting India on its side. The US is concerned that in twenty years' time its traditional allies would become neutral in any confrontation with China, and therefore is cultivating India. Non-proliferation was an irritant in the progress of bilateral relations between the US (or Canada) and India. Only when that issue was sorted out would economic and strategic cooperation flourish. India's integration into the global economy is very important; the deal will promote that.

Ramana agreed that the US was trying to play India off against China, as evidenced by the unprecedented defence cooperation agreement signed between India and the US prior to the nuclear deal. He added that the US has subordinated the NPT to these great power manoeuvres.

Joeck stated that closer US-India ties benefited both countries and it was demeaning to India to assume that it would play the role of a US surrogate in the region. India is also a democracy and will make its own decisions. The rapprochement was not directed against China; rather, it was intended to help India become stronger. The states on the Chinese littoral may appreciate India's enhanced status.

On the question of India's identity as a responsible state, Paul Kapur identified a clearly discernible contrast with Pakistan, which has been shown to be behind several cases of illegal exports of nuclear technology. Moreover, unlike India, Pakistan has used its nuclear weapons for aggressive threat-making. Kapur also observed that while it was theoretically possible for states to learn both negative and positive lessons from India's acceptance by the regime, rogue states by definition were unlikely to become responsible, and so probably would learn only negative lessons.

Neil Joeck agreed that it is important to distinguish states by their behaviour, and contended that the non-proliferation regime's recognition of India as a responsible state was long overdue. North Korea is a criminal state indulging in blackmail and counterfeiting, Iran is a supporter of international terrorism, and Pakistan has the millstone of A.Q. Khan around its neck. These countries should not be equated with India. Huntley asked whether such discrimination should be based on consistent criteria consensually accepted by the global community. Joeck replied that generalized criteria cannot anticipate every contingency, as demonstrated by the NPT's inability to apply uniform criteria for oversight of facilities in NNWS. Hence, President Bush in his February 2004 speech called for a ban on new enrichment and reprocessing facilities.

Regehr asserted the need to take collective decisions on these issues, and suggested that there should be an international process for doing so. Joeck acknowledged that the US had negotiated bilaterally with India. But now, he said, the US government has entered into dialogue with its allies, and the NSG is thinking through the issues.

Stansfield raised the issue of the potential consequences if the new nuclear sharing arrangements falter. The US-India deal, requiring Congressional action, is not guaranteed to go through. A failure could be very damaging. Stansfield noted two scenarios: one, India could become frustrated if its September 2005 vote on Iran does not push forward the agreement with the US; and two, Canada is likely to insist on stiffer conditions than the US before agreeing to uranium exports. So it is important to have a strategy in place to manage the fallout if the US or other countries prove unable or unwilling to cooperate.

Joeck acknowledged that the US Congress would take a lot of convincing, especially since it apparently had been kept out of the loop during the negotiations. Congress is very sensitive on Iran and if India adopts policies at variance with Congressional concerns, Congress may deprive India of some the benefits of the deal. Joeck also raised the concern that worry over the consequences of failure would drive counterproductive compromises. Announcement of the US-India deal has already ended the status quo: making an exception for India to the principle of full-scope safeguards is now on the table. It will be hard to go back, and it is vital to show productive results.

At this point in the discussion, panellists began entertaining questions and comments from others in attendance.

One question addressed the issue of Russia, which had not come up in the discussion up to that point. How would the Moscow-Delhi relationship change as a result of the July agreement? Paul predicted that Russia would be the top bidder for new plants in India. Stansfield added that Russia is currently the only builder of nuclear power plants in India, but that these plants are built on CANDU technology. Hence, the US nuclear industry would have fewer opportunities in the Indian market than Canadian companies.

Gahlaut noted that Russia and France had already been lobbying in the NSG to amend the rules and include India as a member if it agreed to separate the military and civilian parts of its nuclear program—a process termed 'nuclear islanding.' The relationship between Russia and India has changed. The latter is no longer just a consumer. The two are working together on defence projects aimed at new markets.

A follow-up question from the floor asked why India's DAE was seeking Russian cooperation when its *raison d'etre* was to achieve self-sufficiency. Ramana replied that the Indian nuclear establishment had become desperate, yet its hopes that the Koodankulam deal signed with Russia in the 1980s would lead to other such deals had been dashed. Gahlaut added that the DAE had been able to portray itself as under siege as a result of international isolation, thus excusing its inefficiency. The US deal should force the DAE to become more efficient in the face of competition.

Another questioner brought up the issue of the impact of the agreement on non-proliferation in South Asia. What will Pakistan's response be to upgrading of Indian nuclear weapons capability? "Revisionist" is an unfortunate label applied to Pakistan. Following the 1998 tests, statements from VHP leaders and even those like Advani in the BJP were irresponsible. The questioner also expressed his belief that public opinion in India is not as dogmatically pro-nuclear as thought and that no government would lose power if it joined the NPT.

Kapur agreed that the nuclear deal will probably incite Pakistan to ramp up, but asserted that the term "revisionist" was not merely rhetorical, noting that Pakistan and India had very different attitudes to the territorial division of the subcontinent after 1947. He predicted that we will see balancing behaviour from China and Pakistan in reaction to the July agreement.

Paul agreed that Pakistan could not be considered a responsible state because it has allowed the diffusion of nuclear materials. Pakistan's irresponsible behaviour reaches beyond A. Q. Khan's activities; a small Punjabi elite controls its foreign policy. Conversely, Indian Home Minister Advani's threats of hot pursuit against Pakistan-based insurgents, referred to above, were never incorporated into Indian national strategy. Joeck subsequently took up this issue, noting that while we cannot know whether or not A. Q. Khan was a "lone wolf" his fame is a product of his media savvy. Musharraf's government has to be careful because Khan is an icon in the country, not because the government is complicit in Khan's activities.

Ramana further interrogated the notion of responsibility, pointing out that the only time a nuclear weapon was used a NWS was "responsible." Is deterrence itself "responsible"? Responsibility has now come to be defined as the abstention from disseminating nuclear technology, ignoring the irresponsibility of possessing nuclear arms in the first place.

A questioner followed up by observing that Pakistan has agreed to sign the NPT if India does; hence India could be considered a revisionist state, and Pakistan a status quo state, as far as being socialized into the NPT regime's principles is concerned. T.V. Paul agreed that India could be considered a revisionist state in this sense, but not in the classic sense because it is not seeking territorial expansion or other concrete goals. Hence, it is a kind of normative revisionism.

A question from the floor asked what practical implications "NWS" status has and what actual benefits flow to India by being regarded as a NWS? Gahlaut replied that there are no benefits; the status is merely symbolic. But Stansfield observed that there is practical impact: as long as India was not categorized as a NWS, the NSG members could not export to it. Joeck pointed out that "NWS" is a legal invention, but since the NPT is extremely difficult to amend, India cannot become a NWS as defined by the NPT.

Regarding the role of public opinion on nuclear issues in India, T.V. Paul observed that the conclusion of the CTBT in 1996 generated intense nationalism in India. Ramana, on the other hand, stated that in the 1999 election a post-poll survey showed that the majority of people had not even heard of the nuclear tests the previous year. He argued that this shows public opinion does not matter in policy making on the nuclear issue in India, so India could sign the NPT without public opinion being a problem.

An audience member took issue with Ramana's views, asserting that nuclear weapons matter to people in India as symbols of macho nationhood, impelled by a postcolonial consciousness. Ramana responded that nuclear weapons do matter to people, but there is a big difference between what people claim to uphold and what they vote for. He was sceptical of the claims that nuclear issues make any difference in elections and that the government is genuinely responding to public opinion, noting a parallel to how Indian economic policy can diverge from public opinion.

T.V. Paul responded that national security issues do strike a sentimental chord. Political leaders fear that departing from the existing nuclear policy would hand a powerful weapon to their opponents. It is unlikely that an Indian government would survive an opposition onslaught if it were to renounce nuclear weapons and sign the NPT. This is true in part because security policy is more sensitive than economic policy.

An audience member raised a question about India's nuclear cooperation with Iran. Ramana averred that India had a long history of offering nuclear technology to other states such as Vietnam and Iran. Gahlaut stated that the history was not so marked; there were some questionable cases, but today India is minimizing the proliferation threat. Joeck noted that nuclear dealings with Iran come under US law that would apply to Indian entities as India's cooperation with the US opens up.

Each of the invited panellists offered brief concluding comments.

T.V. Paul said that it is only a matter of time before India is accepted as a stakeholder in the non-proliferation regime. The main question is: how fast and in what manner will India be integrated? Human security and development are the issues at the forefront and nuclear power is essential for India's faster economic development.

Paul Kapur concurred that accepting India as a nuclear-armed state is inevitable; India could not be kept out of the regime forever. The danger is that the US may begin thinking that it now "owns" Indian foreign policy. The issue of Iran is an example of the US expecting certain behaviour by India.

Ron Stansfield acknowledged that the current NPT regime is beset by contradictions. There needed to be some method of "normalizing" relations between the non-NPT states and the regime. Stansfield also stressed that, for Canada, the big issue is capping production of fissile materials.

Ross Neil focused on the point that moral suasion would not work to promote nuclear non-proliferation unless the NWS made some progress on Article VI of the NPT. He also noted that the role of economics had so far been trumped by security considerations.

Neil Joeck stressed that the greatest danger facing the world today was the possibility that terrorists would get access to nuclear weapons, and underscored the necessity of prioritizing efforts to meet that threat. One need is to devise enforcement mechanisms within the NPT. The US-India nuclear deal is not everything everyone would wish it to be, but it does advance the core goals of security and non-proliferation.

M.V. Ramana lamented lacuna in the day's discussions, such as the provisions under Article IV of the NPT giving states the right to develop nuclear energy under safeguards. Access to the nuclear fuel cycle allows states to have the potential to construct nuclear weapons; this issue must be addressed.

Seema Gahlaut said that the July deal was not perfect for either the cause of non-proliferation or for India. Yet, the agreement does address important NPT loopholes. The devil will be in the details – the challenge is to come up with the right specifics. She also noted that India will not embrace a fissile material moratorium until China makes a verifiable commitment to ending its production of fissile materials.

Ernie Regehr concluded the discussion period by affirming that we must always deal with the world the way it is. But in that world, the existing non-proliferation regime needs to be built up, not torn down. The NPT does play the role of a disarmament treaty, a large number of states have a stake in this objective, and their interests need to be reflected. One or two countries cannot be arbiters of the common interest. Rules have to apply to all states equally. In this context, new nuclear dealings with India must increase confidence in the regime. Accordingly, the price India is asked to pay should be higher—it should be required to sign the FMCT, as the world cannot allow the indefinite modernization of its arsenal. The obligation to work toward disarmament under Article VI should apply to India as well. It is in the US interest to reinforce these general regime obligations, rather than allow new forms of discrimination based on domestic regime type or short-term security interests.

Conclusion

Nuclear Cooperation with India: Summary and Conclusions

Karthika Sasikumar

On 18 July 2005, US President George W. Bush welcomed Indian Prime Minister Manmohan Singh to the White House with a rare state banquet. Later that day the two leaders issued a Joint Statement declaring that "as a *responsible state with advanced nuclear technology*, India should acquire the same benefits and advantages as other such states" (Office of the Press Secretary 2005) (emphasis added). Many took this to mean that the US had accepted India's self-declared status as a NWS. Just over seven years earlier, after exploding five nuclear devices in the Rajasthan desert, the Indian government had declared "India is now a Nuclear Weapon State". Bush's predecessor, Bill Clinton then described the tests as putting India "at odds with the international community" and "on the wrong side of history" (Clinton 1998). In 2005, however, it seemed that India had emerged on the *right* side of the non-proliferation regime. It had left its undefined and worrisome past behind and was now entering a new era with a validated nuclear identity.

Recognizing the significance of the July 2005 accord for the non-proliferation regime, the Simons Centre invited eight analysts to evaluate the consequences of this acceptance of India as a marginal and informally recognized new entrant to the nuclear club. This conclusion will summarize the main themes emerging in the presentations and discussions, and highlight the key issues and variables conference participants identified in the wake of the agreement.

A Regime Worth Saving? Non-proliferation, Disarmament and Discrimination

While the participants in our conference agreed that controlling nuclear proliferation was a major priority for the international community, not all of them believed that the regime as it stands today was effective in achieving that goal. Seema Gahlaut and T. V. Paul in particular pointed out that the NPT is inherently a *discriminatory* treaty that in effect aims not at complete disarmament, but merely at capping the number of nuclear powers in the race. Thus, by including India in the select nuclear club, the regime was simply recognizing a changed reality—it was not really compromising its basic principles. As Gahlaut pointed out, the critics of the July agreement had not yet proposed an alternative scheme that would accommodate India in the nuclear order. T. V. Paul also emphasized the need for flexibility. India is a rising power and the institutions of the international system will have to amend their rules to accommodate its interests. In fact, Gahlaut added, if India's concerns are not addressed, regressive elements within the country would gain in strength. Moreover, other countries would get the message that only countries that defy the international system and endanger its stability can win the attention of the dominant powers.

The deal would allow India to receive important safety upgrades for its nuclear reactors. It would also facilitate the transfer of safety-related technology for nuclear weapons to India such as the Permissive Action Links that the US has in place to prevent the unauthorized use of nuclear weapons. The deal obliges India to incorporate into its domestic legislation global standards on the transfer of nuclear technology. Thus, it reduces the risk of further horizontal proliferation from India to other countries. Overall, the deal can be said to make India's arsenal safer for the world.

The question remains, however: will the damage in the long run to the regime's credibility as a result of nuclear cooperation with India outweigh this benefit? M. V. Ramana was sceptical that new nuclear trade with India would induce improvements beyond India's current non-proliferation behaviour. Ernie Regehr emphasized that durable non-proliferation solutions require building multilateral consensus through the NPT regime; ad hoc arrangements bypassing the regime will undermine non-proliferation efforts in the long run. All the panellists agreed that the regime has problems. Disagreement centered on where those problems are and how nuclear cooperation with India affects them.

Regional Stability

The overall impact of the deal on nuclear stability in South Asia is, of course, not yet discernible. S. Paul Kapur pointed out that Pakistan would inevitably respond with 'balancing' behaviour to this Indo-US nuclear rapprochement. Pakistan realizes that as a result of this accord, India will be able to receive fuel for its power reactors from abroad. This will free up the existing stockpile for diversion, if India so chooses, to the manufacture of nuclear warheads. Such assumptions may heighten tensions between the two enduring rivals, who have made serious nuclear threats against each other and even gone to war after openly testing nuclear weapons. India and Pakistan have held to their ceasefire since late 2003, however, and may be entering a new period of stability. The dangers of nuclear arms racing are somewhat mitigated in this context.

While it is unlikely that Pakistan would get the same treatment as India, given its currently troubled relationship with the US, its protests against nuclear discrimination would now have more credence. Since it is still improbable that the US will make formal commitments to transfer technology to Pakistan, the country may have to examine other alliances, possibly with China or with other Islamic states. Domestically, the hardliners who want Pakistan to stay out of international regimes will be strengthened.

The repercussions with regards to China are even less predictable. For the last decade, China has been cooperating with the non-proliferation regime that it once castigated. Thus it cannot be too happy about the new legitimization of India's unauthorized entry into the nuclear club. More importantly, China cannot but see India's closer alliance with the US as a potential threat. China could engage in balancing behaviour of its own, with repercussions for American defence planning, for its commitments to Japan and Korea, and for the security of the countries of South-East Asia. Alternatively (or simultaneously), China could pursue with greater fervour the fledgling 'strategic triangle' between India, China and Russia that is often made much of in the Indian media. India and Russia have had a long relationship in the defence and nuclear power sectors, which will only be enhanced by this agreement. China could even support the India-US deal, calculating that India's acceptance of some regime constraints is worth the price of some strategic repercussions.

Certainly, regional dynamics are complex; an enhanced Indian security posture in the region could lead in many different directions. Hence, a US strategy based on the simple idea that bolstering India can balance China could lead to any number of unintended consequences.

Grasping the NWS Identity

In what sense does the India-US accord give India the coveted NWS status? As is well known, the NPT only recognizes five NWS and the procedure for the treaty's amendment was designed to make it an impossible task. So all India could aspire for was recognition by the 'norm leader'—the US—to secure a *de facto* NWS status. The US State Department spokesman Nicholas Burns was careful to clarify: "By taking this decision, we are not recognizing India as a nuclear weapons state. We are simply opening up a channel in order to cooperate on a commercial basis and a technological basis on nuclear power itself and that's a very important distinction..." (Burns 2005). The 'norm leader' of the non-proliferation regime does not support formal NWS status for India. Yet, as Stansfield pointed out, the logic of the US approach to India only holds together if India is named as a NWS. How can one, for example, assert that India is a NNWS while recognizing that it has civilian and military nuclear fuel cycles? How can one make an exemption from the NSG Guidelines for a NNWS?

What is significant is that the 18 July declaration allows India the privileges of a NWS. India will set up a wall of separation between its civilian and military nuclear establishments. It will voluntarily accept full-scope international safeguards to be administered by the IAEA on the civilian portion of its nuclear estate. Unlike the NPT parties who are NNWS, India thus has the right to allow international inspection of those of its facilities that it chooses to designate as civilian. At the same time, India has no real incentives to join the NPT or to cease production of fissile material. Since it is not a party to the NPT it is not bound by the Article VI obligation to work toward disarmament.

It appears as if India has secured the best of both worlds. How did the country manage to leverage itself into this position? Paradoxically India's 1998 tests were the *first* step in achieving a respectable status in the nuclear order. A nuclear test has served as the marker of the ability to manufacture nuclear weapons. This understanding is enshrined in the NPT, which gives NWS status to the five countries that had conducted nuclear tests before 1967. India declared a voluntary moratorium on further nuclear testing and has not since overtly deployed any devices it may already possess. It has also shown measured readiness to accept international controls on its program. India has thus managed to secure a reputation as a 'responsible' nuclear weapons possessor. These policies, combined with India's growing global role and the non-proliferation community's interest in gaining Indian cooperation wherever possible, provided the leverage for India to gain legitimization as a quasi-NWS.

What does this status imply in terms of international norms? In the regime, responsible behaviour has come to be equated with strict controls on the diffusion of nuclear technology outside national boundaries. In stark contrast to China and Pakistan—a contrast that Indian diplomats have avidly played up—India has kept a tight rein on its considerable nuclear expertise. Moreover, India is a democratic state (again unlike its neighbours). Its democratic status leads observers to be more sanguine about the use of its nuclear arsenal purely for deterrence purposes. After the tests, India became more receptive to treaties such as the CTBT that it had rejected in 1996 as discriminatory and hypocritical.

But as M. V. Ramana pointed out, NWS are in fact 'responsible' for holding whole populations hostage in their game of deterrence. From this point of view, describing India as a 'responsible' nuclear power is highly ironic; India has merely embraced the same intentions of the NWS to maintain their nuclear arsenals indefinitely while curtailing further entry to the nuclear club. Here it is worth noting that since 1998 India has turned its back on the radical disarmament rhetoric that had peppered its speeches at international venues.

Nuclear Energy

India has the second-largest population on earth, and its economy has been growing at around 7% for the last few years. Concurrently, the demand for power has risen exponentially. The shortfall in electricity is a severe constraint on productivity and development.

India will now be eligible to receive technology from the US for its nuclear power program. Currently India uses nuclear power for less than 3% of its electricity needs. Extrapolating from the past performance of the AEC, Ramana claimed that we cannot expect India-US cooperation to help satisfy India's growing thirst for power. However, Gahlaut countered with the argument that isolation has contributed to the AEC's inefficiency, or at least has allowed it to make claims to that effect. Only time can tell whether the infusion of new technology will make the AEC more competitive.

Of course, the question remains: is nuclear power the right choice for India's energy crisis? Questions about the safety of nuclear reactors, the disposal of radioactive waste and the cost-efficiency of atomic power plants have plagued the civilian nuclear industry worldwide. On the other hand, questions about the sustainability of a fossil fuel economy have become louder with the unrest in the Middle East. There has been a resurgence of interest in atomic power, also fuelled by concerns about greenhouse gases and climate change. The new generation of nuclear reactors claims to fulfill the 1950s dream of abundant clean energy from the atom (Hannum, Marsh and Stanford 2005). It would be unfair to deny India a chance to participate in the atomic renaissance by blocking its access to the newest technologies available internationally.

Interestingly, the July nuclear deal affects India's energy policy in another way. Since 2004, India has been negotiating with Iran on a natural gas pipeline that would transfer fuel to India's west coast. The 2670 kilometre pipeline would pass through Pakistan. The Bush administration expressed its disappointment at India's cooperation with Iran, a country that it considers part of the 'axis of evil.' Instead, it has been encouraging India to consider an alternative project, partly funded by the oil giant Unocal—the Turkmenistan Afghanistan Pakistan (or TAP) pipeline. American diplomats also made it clear to their Indian counterparts that Congress would approve the July deal only if the Indians were seen as cooperating with the US in its attempt to force Iran to give up its uranium enrichment program. Thus in the crucial vote at the IAEA on 24 September 2005, on the question of referring Iran to the UN Security Council, India broke with the non-aligned countries and voted along with the US. Iran responded immediately, indicating the $21 billion gas pipeline was in jeopardy. This stance has since been softened. However, the question remains: If Iran does withdraw from the pipeline project, will the shortfall in fuel supplies be made up by the US-supported project or by nuclear energy?

Learning from India

Will other countries be encouraged to emulate India? Iran currently occupies a marginal status, barely 'within' the regime. George Perkovich claims that Iran's strategy is to develop a latent nuclear weapons capability without assembling bombs or violating the NPT (Ruppe 2005b). On the diplomatic front, Iranian diplomats are claiming that the US is backing away from the obligations inherent in the NPT's Article IV, which guarantees countries the right to develop atomic power for peaceful purposes. This is similar to India's strategy of critiquing the NWS for disregarding their responsibilities under Article VI that requires them to take steps toward disarmament. Iran is making use of the regime's norms just as India did. To the extent that nuclear cooperation with India bolsters Iran's claims in the eyes of the international community, it will now be much easier for Russia to justify its reactor sales to Teheran.

North Korea, in February 2005, declared that it had manufactured nuclear weapons. Much more of an outcast than Iran, the country withdrew from the NPT and has issued specific threats against the US. Even this flagrant disregard for the regime went mostly unpunished, thanks to divisions among the five NWS. However, Pyongyang has so far been dissuaded from conducting a nuclear test, evidence perhaps of the impact that such a symbolic violation of the regime would have.

While neither country has made significant references to India, clearly Iranian and North Korean diplomats are absorbing the lessons of the Indian experience. Wade Huntley pointed out that this experience holds out both negative and positive lessons. Countries could take away the message that responsible behaviour is rewarded by international acceptance—and refrain from selling their fissile materials or hiring out their nuclear experts. Or, they could learn that the US and by extension, the international community, will accept a *fait accompli*—and be encouraged in their nuclear ambitions. Kapur pointed out that the 'rogue states' about which the international system is concerned are by definition unable to change into democratic and peaceful nations. Thus, states that are at the nuclear 'tipping point' might learn the negative lesson and feel encouraged to cross the nuclear Rubicon with a test.

Bilateral Relations: India and the Sole Superpower

A month before the nuclear deal was signed, India and the US entered into a ten-year defence partnership involving joint weapons production and cooperation on missile defence. While relations between the two countries have been on the upswing for some time now, the Bush administration with its tenuous commitment to multilateral arms control is probably the only interlocutor that the Indians could have found for this deal.

On the Indian side while most commentators welcomed the deal as a victory for Indian foreign policy, some voices were raised in protest. The Left bemoaned the death of India's commitment to disarmament. Defence hardliners worried that the deal would put an end to the ambiguity about fissile material stockpiles that had served India well. Future decisions about the size of the arsenal had been mortgaged to the US, it was alleged. Critics of the deal in India point out that the US is trying to construct a strategic surrogate in Asia. They warn that it may not always be in India's interest to counter China's growing power in the region on behalf of the US. Many commentators were also angered by India's abandonment of its traditional support of Iran. Prime Minister Singh's predecessor, now in the opposition, pointed out that the nuclear deal stops short of recognizing India as a NWS.

Support in the US for India's global power ambitions, rarely in evidence during the Cold War, has facilitated the nuclear deal. Its passage through

Congress, however, is fraught with peril. Sensitivity on the Iran issue has led to reciprocal defensiveness on the Indian side. Congress was also incensed at being kept out of the loop on the negotiations. By playing on anti-China sentiment, the administration could convince Congress that the gains from building up Indian power in Asia outweigh the impact on non-proliferation. India is considering buying $5 billion worth of conventional weapons from the US, including anti-submarine aircraft that it could use in the Indian Ocean to deter Chinese submarines (Linzer 2005). The arms control community in the US has also reacted with scepticism. 16 former senior officials dealing with non-proliferation wrote a letter to Congress alleging that the accord with India would undermine US non-proliferation efforts (Ruppe 2005a).

Canadian Concerns

Two months after the July accord, India and Canada announced the limited resumption of their nuclear cooperation, which had been suspended in 1974. The Canadian government will now permit the sale of dual-use items for use in the civilian sector of the nuclear program in India under NSG guidelines. Canada does not intend to resume the trade in nuclear materials and technology, but even this lower level of cooperation strengthens India's case.

Ross Neil and Ron Stansfield expressed concern at the message conveyed by Canada, an acknowledged leader in non-proliferation. Stansfield pointed out that the deal was by no means a *fait accompli* and that contingency plans for its failure need to be considered. To be able to move forward, the US has to change at least four major pieces of national legislation, including the 1979 NNPA. It has to negotiate a bilateral nuclear cooperation agreement with India, which must be ratified by Congress, and either change the NSG Guidelines or seek a special exemption for India. Should US implementation falter, Canada could be left out on a limb it did not desire to climb in the first place.

Good Nukes, Bad Nukes

Underlying this whole discussion is the question—is it proper to discriminate between countries seeking to acquire WMD? Should the characteristics of the state that successfully obtains them determine the world's response? Neil Joeck answered in the affirmative, saying that it would be unrealistic to treat countries as different as India and North Korea in the same way. The problem with multilateral agreements like the NPT, he continued, was that they were too inflexible to take into consideration the security concerns of a responsible state like India. Ernie Regehr, however, posed a thought-provoking question: if we are to discriminate among different countries, since this is a question of global significance, shouldn't the process be more collective? How can it be proper for the US to usurp the power to make this determination?

Neil Joeck admitted the validity of this question. He predicted that the US would have a difficult time persuading the 40-odd members of the NSG to accept the resumption of nuclear cooperation with India. In the present international climate, the US suffers from a credibility deficit on the issue of preventing WMD proliferation. Some sort of side-payments will have to be offered to other countries that have refrained from nuclear cooperation with India for decades. Regehr pointed out that at the moment, the price for India's admission into the regime is being paid, in the form of an 'opportunity cost', entirely by those countries that were capable of going nuclear but had legally renounced their right to do so in the interest of world peace and stability.

India, insisting on exercising its sovereign right to acquire a military nuclear capability, emerged more or less unscathed by international sanctions. Despite its most egregious defiance of international norms with a nuclear test series in 1998, it has managed to integrate itself into the global nuclear order with the preferred identity of a NWS. Ironically, the nuclear capabilities referenced by those tests have increased global interest in securing Indian non-proliferation cooperation, and so have facilitated India's integration. Its relative restraint after the tests, its accommodation with international standards on technology transfer and its acceptance of discriminatory arms control agreements were obviously influenced by the desire to legitimize its nuclear status. India, therefore, represents both a success and a failure of the nuclear non-proliferation regime.

Background Papers

Canadian Nuclear Cooperation with India in Historical Perspective

Lance Noble

The History of AECL

In 1952 Canada established AECL, the crown corporation responsible for nuclear research. AECL is best known for its development of the CANDU reactor. This is a pressurized heavy water, natural uranium powered reactor. Because the CANDU model produces plutonium it is dual-use by design.

AECL began relations with India in 1956 when Canada provided India with its first nuclear reactor. Because India could ill afford nuclear technology at that time, the Canadian government displayed its commitment to nuclear relations with India by attaching $9.5 million in foreign aid to the $17 million project. As the US provided the 18.9 tonnes of heavy water that the reactor required, it was named the Canada India Reactor US (CIRUS). With proliferation not being considered a serious concern at that time, no enforceable limitations were put on how plutonium derived from the reactor would be used. In fact, the only assurance that Canada did require, "that the reactor and the fissile material it produced would be dedicated solely to peaceful purposes," was placed in a secret annex to the treaty. Significantly, no outside inspections were required (Perkovich 1999).

While Canada and the US trained Indian scientists and shared technical data with them, Indian technicians were working hard to be able to manufacture their own fuel rods after the reactor went critical in 1960. This initiative resulted from the belief that an indigenous fuel fabrication capacity would strengthen India's claim that it was free to use the plutonium that the CIRUS reactor would produce in any way it pleased. India's intentions are suggested by Nehru's 1964 written communiqué to Homi Bhabha, the father of India's nuclear program, that "apart from building power stations and developing electricity there is always a built-in advantage of defence use if the need should arise" (Perkovich 1999).

In 1963 AECL agreed to sell India a 200 MW CANDU reactor. Canada once again signalled its commitment to nuclear relations with India by using the ECIC to finance $35 million of the reactor's total cost of $79 million. While the agreement involved Canada sharing reactor blueprints with India, rising proliferation concerns led Canada to require that Canadian inspectors be given access to the site. As part of the deal India secured the reciprocal right to inspect Canada's Douglas Point reactor. It was also once again agreed that technology provided by Canada would be used for "peaceful purposes." When a deal was struck for a second 200 MW CANDU in 1966, ECIC provided $38.5 million in financing. As the NPT was being negotiated at this time, the subject of inspections was finally brought to the forefront. Instead of bilateral inspections, supervision would now be conducted under the auspices of the IAEA. Furthermore, reactor fuel would now be subject to tight controls.

While India's first CANDU was completed in 1973, the second was still under construction when India used plutonium produced by the CIRUS reactor to conduct its first nuclear test in 1974. As Canada and India had not established the exact limits of the "peaceful purposes" that Canadian technology and materials could be used for, India claimed that it had been a peaceful explosion. Rejecting this logic, Canada's Secretary of State for External Affairs, Mitchell Sharp, stated on 22 May 1974: "The Canadian government has suspended shipments to India of nuclear equipment and material and has instructed AECL, pending clarification of the situation, to suspend its cooperation with India regarding nuclear reactor projects and the more general technological exchange arrangements which it has with the Indian Energy Commission".

In December of that year Canada unveiled a new proliferation policy that required "a binding assurance from buyers that Canadian nuclear material, equipment, and technology would not be used for a nuclear explosive device, and rejected the excuse of 'peaceful nuclear explosions'" (Martin 1996). It also proscribed the use of technology provided by Canada for nuclear tests and ruled that nuclear technology exports would now take place under the rubric of the Department of Industry, Trade and Commerce and the Atomic Energy Board. Canada formally brought nuclear cooperation with India to an end on 18 May 1976 after India refused to agree to subject reactors other than the two CANDUs to inspections. This caused considerable difficulties for India's nuclear program as it was still fairly dependent on Canadian assistance. This can be seen in the longer delays and costs involved in completing nuclear reactors and heavy water plants. The second CANDU was only completed in 1981.

Possible Future Canada-India Nuclear Cooperation

Before AECL can begin to resume sales to India, Canada and India must sign a new nuclear cooperation agreement. This trade treaty would involve IAEA safeguards, and reports on the production of nuclear fuel and plutonium. It would also require India to decouple its civilian and military programs, with the civilian program becoming subject to IAEA inspections.

To some extent the 26 September 2005 policy announcement by Canada's Minister of Foreign Affairs, Pierre Pettigrew, is a re-articulation of a longstanding offer that Canada made India in the early nineties to improve safety assistance. But India has only recently become willing to meet the condition that its civilian program be subject to inspections. The fact that Canada is raising the possibility of selling reactors highlights the trend toward opening trade in peripheral areas that are dual-use.

A survey of AECL's nuclear reactors makes it clear that Canada possesses technology that would be beneficial to India's civilian and military nuclear programs. While India has a total of 13 completed reactors based on CANDU technology and 5 more currently under construction, 16 of these have a gross capacity of 220 MW or lower, the final two having a net capacity of 540 MW. By way of comparison, AECL's CANDU-6 and CANDU-9 models have a capacity of 700 MW and 900 MW respectively. Known as CANDU ACR reactors, they represent the latest advances in nuclear technology; they are one third smaller yet maintain the same power capacity as their predecessors, and use one fourth of the heavy water. Additionally, by using slightly enriched uranium, the ACR reactors' fuel lasts three times as long as that required by reactors using natural uranium.

The possibility of selling these advanced reactors to India highlights five areas of concern with regards to Canada's interest in upholding the NPT and the manner in which AECL operates.

First is the need to establish a firewall between India's civilian and military nuclear programs. As this decoupling will be a difficult process, the recent US-Indian agreement stated that "the identification and separation of civilian and military facilities and programmes in India will not occur precipitously, but in a phased manner." Furthermore, as the Additional Protocol that India will need to sign with the IAEA will likely be similar to the agreements that the IAEA has with the five recognized nuclear powers, India will be allowed to exclude military facilities and even areas of civilian facilities on the grounds of national security.

New Delhi has stated that India will reserve the right to switch a reactor's classification from civilian to military, leaving the possibility that a new generation of more advanced Canadian reactors would one day be integrated into India's military program. India's refusal to cease the production of fissile material until a FMCT is signed (unlikely in the short-term) and reports that India is experimenting with low-yield weapons underscore the seriousness of such potential conversion.

The second concern is the amount of information that is being shared with India. A trend toward increased information sharing began to emerge in 1988 as India and Pakistan were permitted to join the COG. This is a group of countries that own CANDU reactors that share technical and safety information. India and Pakistan's entry into the COG was primarily motivated by safety concerns. A meltdown in a poorly maintained CANDU reactor would be environmentally disastrous but the impact of such an event on the CANDU brand and future sales was also a consideration.

India has noticed that there is a great deal of friction between COG and DFAIT with regards to how much information COG can share with India, a disagreement that India has not been above utilizing for its own benefit. As a result, DFAIT has had to brief COG about what information it is allowed to share. Communication between Canadian and Indian scientists working in the field complicates the matter as it is considerably more difficult to enforce export controls on the expertise of individual scientists. This challenge is highlighted by reports after the Indian nuclear tests in 1998 that Indian scientists had been able to maintain access to Canadian expertise through personal contacts.

The third concern is the manner in which AECL projects are funded, which brings into question whether reactor sales to India are in the interests of Canadian taxpayers. The case of AECL's 1996 sale to China of two CANDU reactors is telling. The Federal government used the Canada account of EDC, the successor to the ECIC, to make a $1.5 billion loan guarantee to the state bank of China. Federal documents released in relation to a case brought to court by the Sierra Club of Canada confirmed that if the Chinese had defaulted on the loan Canadian taxpayers would have been held financially responsible. As the Canadian government has refused to release financial information related to the project it is impossible to discern the profit/loss and the level of risk it involved. The government states that releasing this information would run against the national interest as it could damage the possibility of future CANDU sales. The fact that Canada was willing to make a similar deal with Turkey and that five of the eight CANDU reactors that have been sold have been funded by the EDC makes it likely that future sales to India will take the same form. Unless it can be shown that these sales will be profitable the obvious question of why Canadian citizens should want to make them must be asked.

The fourth concern is the manner in which contracts are won. Due to past corruption AECL was required to declare the names of its overseas agents. In 1998 AECL unilaterally and without prior warning decided to cease these declarations, claiming that this change was necessitated by the risk that their agents would be bribed by their competitors. AECL's past record of handling public funds and its lack of accountability bring into doubt its ability to conduct future negotiations with India.

The fifth and final concern is the amount of funding that AECL has received from the Federal government. In addition to subsidies of $16.6 billion from 1953 to 2000 (in 2000 dollars) AECL continues to receive $100 million a year. By way of contrast, renewable energy received $12 million in 2000 (Martin 2000). In spite of this generous funding AECL has not filed corporate plans to Parliament since 1995. One observer has claimed that "The CANDU project should have been declared a commercial failure and wound up two decades ago" (Adams 2002).

As no new reactors have been constructed in Canada since 1992, it is understandable that AECL finds India, with its talk of adding 25,000-30,000 MW of capacity over the next twenty-five years, to be an attractive market. With that said, AECL's ability to turn government subsidization into commercial success has been limited. After setting a sales goal of "ten reactors in ten years" in 1995, AECL only managed to sell two by 2005. While AECL's strategy seems to be to get its foot in the door in the hopes of additional sales in the future, most importing nations have preferred to manufacture their own reactors after making one initial purchase. AECL's attempts to open export markets that may pose security risks while continuing to receive hundreds of millions of taxpayers' dollars raises questions as to whether AECL's operations serve the Canadian national interest.

Nuclear Cooperation with India and its Consequences for the NPT Regime

Ariel Zellman

On 26 September 2005, the Canadian government issued a statement signalling its intention to formally abandon its seven-year self-imposed moratorium on nuclear cooperation with India. The statement expressed Canada's willingness to resume trade in dual-use nuclear technologies (those which have both civilian and military applications). This is a radical departure from Canadian nuclear non-proliferation policy established following the nuclear weapons tests by India and Pakistan in 1998. As a party to the NPT, Canada's offer to resume nuclear cooperation stands in contrast to its stated commitment to prevent the spread of nuclear technology to states in which there is an unacceptable risk that it may used to develop weapons. It is the goal of this paper to lay out how the Canadian decision came about, the basics of the NPT regime as they apply to this case, and the ramifications of new nuclear cooperation with India for the NPT and the broader norm of nuclear non-proliferation.

The US-India Nuclear Deal

Canada's decision was not made in a vacuum. Following the 18 July 2005 meeting between US President George W. Bush and Indian Prime Minister Manmohan Singh in Washington, the two leaders hailed an "emerging global partnership" and agreed on a number of bilateral issues including the transfer of American nuclear technology and materials to India for the development of nuclear energy. Now, other Western states that have also had moratoriums on nuclear cooperation with India are beginning to abandon them. The UK and France have signed similar deals, signalling a tacit acceptance of India's NWS status (Boese 2005). While the US Undersecretary of State for Political Affairs Nicholas Burns told reporters on 19 July 2005, "By taking this decision, we are not recognizing India as a nuclear-weapon state," Prime Minister Singh assured the Indian Parliament on 29 July 2005, "There is nothing in the Joint Statement that amounts to limiting or inhibiting our strategic nuclear weapons program."

The Bush administration has demanded that India separate its military and civilian programs and voluntarily abide by IAEA safeguards. India, in turn, has agreed to continue a self-imposed moratorium on testing, ensure tight export controls, and participate in the negotiation of a FMCT. But the concern that dual-use technologies will be used inappropriately remains. To implement the agreement, the White House must win the approval of both the US Congress and international bodies that restrict the sale of sensitive nuclear technologies. On 26 October 2005, at the second hearing on the issue, four out of the five witnesses empanelled by the US House Committee on International Relations affirmed the conventional wisdom that such a deal weakens non-proliferation rather than strengthening it.

Of further concern is that India has been generally supportive of Iran in the latter's quest to legitimate its nuclear energy program. Members of the US House of Representatives International Relations Committee have been intensely critical of the deal with India in this regard, as the US is pressing Iran and North Korea to abandon their nuclear programs. One particularly outspoken congressman, Edward J. Markey (D-Mass.), condemned the agreement as a "dangerous proposition and bad non-proliferation policy" saying that it undermines American credibility and previous efforts to limit nuclear weapons proliferation (Milbank and Linzer 2005). It is significant to note, however, that India voted with the US on an IAEA resolution which requires that Iran be reported to the UN Security Council for its illicit nuclear activities (Nayar 2005). Whether India's decision was more about securing the details of the joint nuclear deal than a desire to take firm action against Iran remains to be seen.

The marked foreign policy departure represented by the US initiative in part demonstrates an ongoing effort by the US to court India, the world's most populous democracy and a constant critic of American foreign policy as the leader of the Cold War era Non-Aligned Movement. As Sumit Ganguly has noted:

From the outset [the Bush] administration has granted India the prominence it seeks (and arguably deserves). Indeed, even before Bush was elected, his principal foreign policy adviser Condoleezza Rice was noting in the pages of *Foreign Affairs* India's rise as a regional power. Once in office, the Bush team quickened the pace of bilateral military-to-military contacts, agreed to a modest but critical number of weapons sales, and decided that relations with India would no longer be held hostage to Pakistani misgivings and objections (Ganguly 2005).

Nicholas Burns confirmed the importance of such investment by characterizing the new deal as "a major move forward for the US" and "the high-water mark of US-India relations since 1947" (Milbank and Linzer 2005). Furthermore, this move is clearly part of a much larger push to bolster India's power and influence in Asia as a counterweight to the emerging economic and military power of China, an intention privately confirmed by several officials in the administration. Pentagon officials disclosed just two days after the US-India agreement was signed that "they expected India to start purchasing as much as $5 billion worth of conventional military equipment as a result of the deal," including anti-submarine patrol aircraft to spot Chinese submarines in the Indian Ocean and Aegis radar for Indian destroyers in the Straits of Malacca to monitor Chinese naval activities. Coupled with the Pentagon's simultaneous release of an assessment of China's military strength which notes that Beijing is increasing its nuclear arsenal and its capability to target India, Russia and "virtually all of the US," the US objective to support India in order to maintain a military balance in the region appears undeniable (Linzer 2005).

The NPT

The NPT stipulates that the only states that may possess nuclear weapons are those which detonated nuclear devices prior to 1 January 1967: the US, Russia (as the successor state to the USSR), the UK, France, and China. Although France and China did not become signatories until 1992, their eventual accession to the treaty was understood to be dependent on the acknowledgement of their status as NWS. All other states, however, were designated NNWS, and by signing, they effectively renounced the potential to ever manufacture or possess such weapons. Nuclear threshold states were therefore compelled to either conform to a non-nuclear status or reject the treaty entirely. India chose the latter course. As the ultimate goal of the treaty was to achieve "general and complete disarmament under strict and effective international control," the idea that additional NWS could ever be recognized was anathema to the "spirit" of the NPT and was therefore not considered.

Understanding that many states would not agree to limit their own military capabilities without significant inducements from the privileged five, the NPT lays out both the specific responsibilities for both classes of states and the benefits to be enjoyed by NNWS signatories. Consequently, the NPT maintains a dual focus of preventing the proliferation of nuclear weapons and facilitating the development of atomic energy for peaceful purposes. The NWS undertake not to transfer, assist, encourage, or induce NNWS to develop, attain or control such arms, while it is the responsibility of the NNWS not to receive, manufacture, acquire, or seek assistance in developing said weapons. Furthermore, it is incumbent on NNWS to accept safeguards imposed by the IAEA on all fissionable materials to ensure appropriate civilian use and prevent illegal transfers. No state party, even a NWS, is allowed to transfer any fissionable material or materials used for nuclear processing to any NNWS unless the material is subject to IAEA safeguards. In return, all NNWS signatories are entitled to the benefits of research provided by nuclear explosions conducted by NWS and are guaranteed the "inalienable right ... to develop research, production, and use of nuclear energy for peaceful purposes" within limitations spelled out by the treaty.

NPT Consequences of Nuclear Cooperation with India

The challenge now posed to the NPT is that the opening of nuclear trade relations with India legitimizes India as a NWS. These deals accord India a high degree of respectability, abandoning economic isolation as a means to challenge its development of a nuclear weapons program. India also will enjoy the benefits of trade as a NNWS without the capacity of legal enforcement for violations of safeguards that would be imposed by the treaty on NWS.

In order to deal with this complication, the potential exporters have decided to work around these legal obstacles and encourage India to voluntarily adhere to non-proliferation norms without necessitating its disarmament. One instrument for controlling India's activities in this way is careful multilateral monitoring of the technologies it purchases through the NSG. Formed following India's 1974 nuclear detonation, the group currently consists of 44 states, including Canada and the US. The NSG has a trigger list for a number of nuclear-related technologies and materials that are, in turn, subject to IAEA safeguards. This measure imposed by the NSG ensures that no member will export any item on the trigger list without the importing country subjecting itself to particular IAEA monitoring.

In 1992, the NSG adopted further controls specific to the transfer of dual-use technology. The Dual-Use Guidelines prohibit the transfer of any controlled item to a NNWS if there is an unacceptable risk that it may be used for military applications (Anon nd). The recipient of these goods must report specifically how the materials will be used, guarantee that they will not be used for prohibited activities, and receive the suppliers consent before the transfer of any item so received to another state. The exporting state maintains control over all technical assistance and information for development, production, and use of the dual-use technology, thus encouraging accountability by the exporting state and theoretically preventing the recipient state from illicitly using the acquired goods.

These NSG guidelines are one of the most significant roadblocks in the way of both the American and Canadian nuclear cooperation deals. However, since these guidelines are legally binding on neither the recipient nor the supplier, and many NSG members are rushing to accommodate India's civilian nuclear ambitions, the NSG may prove not to be the best means to ensure continued respect for non-proliferation norms.

Another control mechanism is the IAEA Model Additional Protocol, drafted by the agency in 1997 to encourage non-nuclear weapons signatories of the NPT to adhere to stricter safeguards and inspections over nuclear facilities and materials. This agreement extends reporting requirements beyond accounting for fissionable materials and technology required to utilize them, to include the use and development of those technologies not directly involved in fission such as information on uranium mines, thorium plants, and information required on materials monitored by the NSG (Anon 2005a). The Additional Protocol also expands the number and type of facilities which the agency can inspect and streamlines the visa application process to facilitate inspections, and allows IAEA agents to sample soil at both declared and undeclared sites and over a wider area than previously permitted (Kimball 2005).

While more comprehensive than the original IAEA requirements, the new terms are entirely voluntary. 106 States have signed Additional Protocols as of 25 November 2005; however, they are only in force in 69 of these countries. Protocols have been completed with Iran and Libya but they have yet to go into force (Anon 2005d). Furthermore, as of 17 October 2005, Iran has suspended its implementation of the Additional Protocol until it has succeeded in "obtaining recognition of its right to complete the nuclear fuel cycle" (Anon 2005c). India has suggested that it may accept these guidelines, but would likely apply them selectively to designated *civilian* nuclear plants. Since the deals made by Canada, the US, and others are contingent on a civilian-military split, they are not in a position to dictate similar safeguards and monitoring of Indian military facilities.

By opening the door to India both as a NWS and as a legitimate recipient of nuclear technical assistance, the underlying concepts behind the rules of the NPT have been broken, and may encourage other states in similar non-compliance to seek comparable accommodations. Already Pakistan, according to some sources, may be seeking a similar deal with the US, France, and the UK as obtained by India (Anon 2005f). Having renewed nuclear cooperation with India, it will be difficult for these countries to maintain neutrality between the two competing powers and deny Pakistan such a deal – even though the country has been an egregious nuclear proliferator, selling technology to Libya, Iran, and North Korea (Cronin 2005). Furthermore, it is troubling to conceive of the effect this deal might have on states which have previously given up their nuclear weapons programs, such as South Africa, Argentina, Brazil, and Egypt. Although it certainly will not lead to the acquisition of weapons by these states, it may persuade them to rethink their stance on non-proliferation.

Whatever the final results are of the nuclear cooperation agreements that have been made with India thus far, they present a substantial challenge to the continuity of the NPT regime in their tacit recognition of the legitimacy of a non-signatory's possession of a nuclear arsenal. If anything, these agreements highlight the problems inherent in the inflexibility of the treaty. A state such as India, which seems unlikely to renounce its weaponised status in the near future, cannot be formally incorporated into materials control structure. These agreements also potentially produce new loopholes by which states of concern such as Iran and North Korea could have grounds for remaining party to the treaty while actively undermining its provisions.

Whether the end result is the troubled maintenance of the status quo, abandonment of the NPT, or revision of the NPT to allow for formal dealings with unrecognized NWS, the standard by which nuclear non-proliferation has been judged and managed over the past fifty years is undoubtedly being challenged. Whatever the outcome, one must continue to hope that the international community is able to come to some agreement by which those who develop and maintain nuclear technological capabilities continue to be held morally responsible and legally accountable for their conduct.

India and the NPT: Power, Legitimacy and Legality

Michael Cohen

This paper considers the assumptions about the role of power underlying the two positions that characterize the debate regarding the implications for the NPT of nuclear cooperation with India.

The first position applauds the new initiative to India as a realist recognition of emerging realities. In this view, the dilution or even demise of the NPT is simply an adaptation to transforming power dynamics. This position considers power to be brute military and economic might and downplays the role of legitimacy. The second position focuses on concerns that new nuclear cooperation with India will bolster the perceived connection between nuclear weapons possession and international status. In this view, the NPT continues to play a valuable role in stemming nuclear non-proliferation; accepting India as a bona fide NWS would undermine those efforts. This position thus stresses the need for legitimacy to transform military and economic might into real power. Both of these interpretations have validity. Yet debate has become polarized. The two positions seem irreconcilable only because the assumptions about power and legitimacy underlying them are seen to contradict one another.

They do not. Both these sets of arguments marginalize the 'power of rules.' This factor, rather than being epiphenomenal, actually serves to gird both material *and* normative power structures. Hence, a more useful understanding of underlying power dynamics involves acknowledging the centrality of both force and legitimacy to power, bolstered by a robust legal framework. The "power of rules" viewpoint helps explain the prevailing situation with respect to nuclear cooperation with India and offers a basis for tenable policy proposals that reconcile concerns over both power and legitimacy.

Such an understanding can help bridge current contentions over how nuclear cooperation with India will impact the NPT and help chart a viable future course for effective global non-proliferation cooperation. Indeed, some carefully crafted proposals for coping with contemporary non-proliferation challenges, reviewed below, represent attempts to reach the "middle ground" precisely by emphasizing the role of rules in linking power and legitimacy. Such approaches are the best means to maintain the future viability of the nuclear non-proliferation regime.

The Power of Force

In July 2005 the Bush Administration dubbed India "a responsible state with advanced nuclear technology" (Kronstadt 2005). This position moves away from five decades of attempts to delegitimize India's nuclear acquisition. The Bush Administration's approach sees forces of legitimacy as holding little practical utility, and places its faith in maximising US military strength rather than in establishing international law or international norms (Goldstein 2004). As the value of the NPT is viewed in instrumental terms and as the benefits have come to be outweighed by its costs, its viability has seemed to diminish.

In this view, the costs and benefits of the NPT were both the cause of its success from the 1970s to the 1990s and the cause of its being at a crossroads today (Huntley 2005). Thus, the Bush administration views the NPT as forcing US military interest into a strategically deleterious position. The NPT is perceived not to fit the new strategic reality and is correspondingly relegated to the margins of the policy toolbox. Indeed, the change in US nuclear policy toward India has already incited Ottawa and a number of other governments to similarly reconsider their policies toward India and the NPT.

This position exemplifies the arguments of Kenneth Waltz as to the strategic impact of possessing a nuclear capability (Sagan and Waltz 2003). In this view, nuclear weapons constitute the ultimate deterrent. While not containing all forms of armed violence, their possession encourages parties to a conflict to ensure that minor skirmishes do not become major entanglements. Through Waltzian reasoning, possession of nuclear weapons can be (albeit to differing degrees) in the interest of any state.

From this viewpoint, the NPT forces most states to eschew useful capabilities, cementing a strategically advantageous position for the US because it was permitted to retain its nuclear weapons – at least over the medium term. As Deborah Ozga has pointed out, this arrangement served US security interests by "erecting significant barriers to nuclear proliferation without forcing it to make some difficult sacrifices in terms of disarmament and inspection" (Ozga 2000).

The Power of Legitimacy

Certain states in the non-nuclear weapon majority have reacted to these initiatives to India with skepticism. They are less motivated to fulfill their own treaty obligations and even less motivated to participate in efforts to strengthen the regime. Already Brazil has refused to permit IAEA access to certain parts of a new uranium-enrichment facility (Goldstein 2004). Defending the need for the new plant, Brazilian energy adviser Rogerio Cezar Cerqueira Leite stated, "Without enriched uranium, you don't have nuclear technology... It's not just national prestige. If you don't make it yourself, you will always be behind in the nuclear race" (Chang 2006).

To critics of new nuclear cooperation with India, such reactions epitomize the risks of ignoring the NPT's legitimization role. For observers such as Daryl Kimball, the main source of the success of the NPT is its widespread legitimacy, which creates strong incentives to abide by its provisions even when they conflict with more narrowly defined self-interest.   It is not that power does not matter, but that military and economic might are not *all* that matter to power.   While the Bush Administration's dismissal of this function leads it to undervalue the NPT, observers such as Kimball point to further proliferation emerging from the NPT's deterioration and stress that its legal integrity must be upheld. In former US Deputy Secretary of State Strobe Talbott's words, the adjustment of the legal infrastructure entailed in reopening nuclear trade with New Delhi will cause "other states to regard the NPT as an anachronism, reconsider their nuclear self-restraint, and be tempted by the precedent that India has successfully established and that now, in effect, has an American blessing" (Kimball 2005).

India's formal display of a nuclear capability in May 1998 posed an immediate challenge to the foundations of the NPT. Not only did other NNWS have to cope with a nuclear India, leading to increased incentives for horizontal proliferation, but the impact of the failure to bring India into the enforcement mechanisms and legal limits of the treaty became palpable. India's nuclear acquisition thus dealt a blow to the legitimacy of the NPT. The Bush Administration extended its offer to India because it perceived that US interests regarding India extended beyond non-proliferation. This realist position does not so much deny the contentions of those who see the nuclear deal as undermining the legitimacy of the NPT – it simply dismisses such contentions as inconsequential.

The Power of Rules

Kimball is right to allude to the coming crisis of legitimacy of the NPT, brought about more generally by US disinterest and propelled by the offer to supply nuclear materials to India (Kimball 2005). His analysis, however, marginalizes the pervasive military and economic exigencies that made the offer such a strategic necessity in the first place. It is here that the Bush Administration's thinking is sound. But both positions misunderstand the legal function of the NPT; while the policy from Washington marginalizes the interests of the non-nuclear states at the expense of the five nuclear states, the Kimball position marginalizes the latter at the expense of the former. The NPT and the broader non-proliferation regime can only be sustained by serving the interests of all parties. Strengthening the NPT in this manner will be good for nuclear non-proliferation; pursuing this goal could be a good policy position for Canada.

Analysts from the Carnegie Endowment for International Peace (CEIP) have made one such attempt. Their report recognizes these complex dynamics and proposes a new blueprint to the international nuclear non-proliferation regime (Perkovich, Mathews, Cirincione, Gotemoeller and Wolfsthal 2005). Stressing the need for a legal framework that balances the interests of the US and other nuclear states with the non-nuclear states, it calls for (among other recommendations) a "toughened regime that, through precluding acquisition of uranium enrichment and plutonium reprocessing plants by any additional state, provides, in return international guaranteed, *economically attractive* supplies of the fuel services necessary to meet nuclear demands" (italics added).

Another initiative was put forward by the IAEA Director-General. Mohammed El Baradei's endorsement of the India-US deal shows the basis of his philosophy: "making advanced civil nuclear technology available to all countries will contribute to the enhancement of nuclear safety and security" (El Baradei 2005a). The lack of progress by the NWS toward disarmament allows the NNWS signatories to the NPT to renege on their non-proliferation commitments. Such states could conceivably produce nuclear weapons within months of withdrawing from the treaty. An assurance of supply of civilian nuclear materials to the NNWS, many of which face nuclear energy crises, could alleviate their incentives to acquire nuclear materials outside the treaty parameters. Venezuela's government has recently asked Argentina about the possibility of providing technical expertise to help develop nuclear energy for peaceful purposes (Serrat 2005). El Baradei has recently emphasised the importance of this assurance of supply: "by providing reliable access to reactors and fuel at competitive market prices, we remove the incentive or justification for countries to develop indigenous fuel cycle capabilities. In so doing, we could go a long way toward addressing current concerns about the dissemination of such capabilities" (El Baradei 2005b).

These approaches to handling advanced civil nuclear technology can satisfy the signatories' nuclear energy needs and thus legitimize the legal framework of the NPT amidst the changing dynamics that increase the benefits from proliferation. This would ensure the longevity of the NPT as a viable non-proliferation tool. This guarantee is based on expanded access to nuclear facilities under the 1997 additional protocol that enables the IAEA to identify possible undeclared activities. El Baradei has insisted that India identify and place all its civilian nuclear facilities under IAEA safeguards and sign and adhere to a further additional protocol with respect to civilian nuclear facilities (El Baradei 2005a). New Delhi's progress in this area will be a key indicator of success for this approach.

Significant conflicting political interests will complicate a solution to the collective action problem of fuel cycle internationalization. However, the CEIP and IAEA proposals are steps in the right direction. Failing significant progress toward internationalization of the nuclear fuel cycle, Canadian nuclear cooperation with India should be subject to the latter's adherence to certain provisions of the NPT.

The success of the NPT has been, and always will be, its ability to serve the interests of both the nuclear powerful and the nuclear powerless. But as the transforming interests of the former challenge the integrity of the NPT, the latter will be hard pressed not to push for similar transformative changes. These will likely take the form of nuclear proliferation. But meeting this problem by incorporating fuel cycle internationalization into the NPT would be seen by many states as a new restriction on civilian nuclear development, eroding perceptions of the intrinsic benefits that the treaty offers. The sooner solutions to current proliferation challenges can be brought within the legal confines of the NPT, the better it is for nuclear non-proliferation.

The Canadian Government confronts new dilemmas for its own policy: seeking viable responses to the changed conditions created by the US initiative while remaining faithful to Canada's own non-proliferation commitments in the context of its own troubled history with India. Proposals such as those offered by the CEIP and IAEA, evoking the 'power of rules', can offer guidance. An internationalized fuel cycle under IAEA control that also provides increased access to peaceful nuclear energy could satisfy the non-nuclear states, discourage the majority of them from proliferation, and thus maintain the viability of the treaty. Ottawa could find a resolution of its conflicting imperatives in taking a leading role in promoting such an initiative.

Bibliographies and Biographies

References

Abraham, Itty. 1997. Science and Secrecy in the Making of Postcolonial State. *Economic and Political Weekly*:2136-46.

Adams, Tom. 2002. Last Call for AECL Subsidies. *National Post*, 20 March.

Anon. 2000. *NPT RevCon 2000, Final Document* Carnegie Endowment for International Peace, 2000 [accessed 11 March 2006]. Available from http://www.ceip.org/programs/npp/NPT2000FinalText.htm.

Anon. 2005a. *The IAEA 1997 Additional Protocols at a Glance* Arms Control Associations, 2005a [accessed 11 March 2006]. Available from http://www.armscontrol.org/pdf/iaea1997additionalprotocolataglance.pdf.

Anon. 2005b. India, Iran and the Congressional hearings on the Indo-U.S. nuclear deal. *Hindu*, 1 October.

Anon. 2005c. *Iran suspends NPT additional protocol until IAEA recognizes its rights* Arabic News, 2005c [accessed. Available from http://www.arabicnews.com/ansub/Daily/Day/051017/2005101737.html.

Anon. 2005d. *More states sign safeguards agreements and additional protocols* IAEA, 28 November 2005d [accessed 11 March 2006]. Available from http://www.iaea.org/NewsCenter/News/2005/safeguardsrights.html.

Anon. 2005e. *N-ties with India unique to us: America* Express India, 20 October 2005e [accessed 11 March 2006]. Available from http://www.expressindia.com/fullstory.php?newsid=56932.

Anon. 2005f. Pakistan seeks N-deal similar to India. *Dawn*, 4 October.

Anon. nd. *Nuclear Suppliers Group* Federation of American Societies, nd [accessed.

Boese, Wade. 2005. Bush Promises India Nuclear Cooperation. *Arms Control Today*.

Burns, R.Nicholas. 2005. *Briefing on the Signing of the Global Partnership Agreement Between the United States and India* US State Department, 19 July 2005 [accessed 17 November 2005]. Available from http://www.state.gov/p/us/rm/2005/49831.htm.

Carter, Ashton B. 2005. *The India Deal: Looking at the Big Picture* 2 November 2005 [accessed 3 January 2006]. Available from http://bcsia.ksg.harvard.edu/publication.cfm?ctype=testimony&item_id =51.

Central Intelligence Agency. 2004. *Rank Order--Purchasing Power Parity* 1 November 2004 [accessed 11 March 2006]. Available from http://www.cia.gov/cia/publications/factbook/rankorder/2001rank.htm l.

Chanda, Nayan. 1999. The Perils of Power. *Far Eastern Economic Review*, 4 February.

Chang, Jack. 2006. *Brazil Takes a Major Nuclear Step* San Jose Mercury News, 12 February 2006 [accessed 11 March 2006]. Available from http://www.mercurynews.com/mld/mercurynews/news/world/1385415 6.htm.

Clinton, William. 1998. *President's Radio Address* 25 May 1998 [accessed 1 November 2005].

Cronin, Richard P. 2005. *Pakistan's Nuclear Proliferation Activities and the Recommendations of the 9/11 Commission: U.S. Policy Constraints and Options* Congressional Research Service, 25 January 2005 [accessed 11 March 2006]. Available from http://www.fas.org/spp/starwars/crs/RL32745.pdf.

Dias, Xavier. 2005. DAE's Gambit. *Economic and Political Weekly* 40 (32):3567-69.

Einhorn, Robert. 2005. *Should the US Sell Nuclear Technology to India? - Part I* Yale Global Online, 8 November 2005 [accessed 11 March 2006]. Available from http://yaleglobal.yale.edu/display.article?id=6474.

El Baradei, Muhammad. 2005a. *IAEA Director General Reacts to US-India Cooperation Agreement* International Atomic Energy Agency, 20 July 2005a [accessed 11 March 2006]. Available from www.iaea.org/NewsCenter/PressReleases/2005/prn200504.html.

El Baradei, Muhammad. 2005b. *Reflections On Nuclear Challenges Today* International Atomic Energy Agency, 6 December 2005b [accessed 11 March 2006]. Available from http://www.iaea.org/NewsCenter/Statements/2005/ebsp2005n019.html.

Ganguly, Sumit. 2005. *Giving India a pass* Foreign Affairs, August 17 2005 [accessed 1 March 2006]. Available from http://www.foreignaffairs.org/20050817faupdate84577/sumit-ganguly/giving-india-a-pass.html.

Goldstein, Ritt. 2004. This Nuclear Age. *Asia Times*, 17 May.

Hannum, William H., Gerald E. Marsh and George S. Stanford. 2005. Smarter Use of Nuclear Waste. *Scientific American* 293 (6):84-91.

Hart, David. 1983. *Nuclear Power in India: A Comparative Analysis*. London: George Allen and Unwin.

Huntley, Wade. 1999. Alternate Futures After the South Asian Nuclear Tests: Pokhran as Prologue. *Asian Survey* 39 (3):504-24.

Huntley, Wade. 2005. *The NPT at a Crossroads* Foreign Policy in Focus, 1 July 2005 [accessed 8 March 2006]. Available from http://www.fpif.org/fpiftxt/144.

Karat, Prakash. 2005. Betrayal on Iran. *Indian Express*, 30 September.

Kimball, Daryl G. 2005. *A Nonproliferation Reality Check* Economic and Political Weekly, 27 August 2005 [accessed 11 November 2005]. Available from http://www.epw.org.in/showArticles.php?root=2005&leaf=08&filename =9037&filetype=html.

Kronstadt, K. Alan. 2005. *India-US Relations* Congressional Research Service, 15 November 2005 [accessed 8 March 2006]. Available from http://www.fpc.state.gov/documents/organization/57185.pdf.

Linzer, Dafna. 2005. Bush Officials Defend India Nuclear Deal. *Washington Post*, July 20, A17.

Lodgaard, Sverre. nd. The Nuclear Deal with India. In *Norwegian Institute of International Affairs*.

Markey, Edward. 2005. *House Energy Conference Committee questions logic of new India nuke strategy* News from Ed Markey, 19 July 2005 [accessed 11 March 2006]. Available from http://markey.house.gov/index.php?option=com_content&task=view&i d=560&Itemid=56.

Martin, David. 1996. Exporting Disaster: The Cost of Selling CANDU Reactors. Ottawa: Campaign for Nuclear Phaseout.

Martin, David. 2000. *Nuclear subsidies to AECL total $12 million* Campaign for Nuclear Phaseout, 21 November 2000 [accessed 11 March 2006]. Available from http://www.cnp.ca/media/nuclear-subsidies-11-00.html.

McGoldrick, Fred, Harold Bengelsdorf and Lawrence Scheinman. 2005. The US-India Nuclear Deal: Taking Stock. *Arms Control Today.*

Mian, Zia and M. V. Ramana. 2005. *Feeding the Nuclear Fire* Economic and Political Weekly, 27 August 2005 [accessed 2005 14 November]. Available from http://www.fpif.org/fpiftxt/659.

Milbank, Dana and Dafna Linzer. 2005. US, India May Share Nuclear Technology. *Washington Post*, 19 July, A01.

Nayar, K.P. 2005. No-choice Delhi votes with US. *Telegraph*, 25 September.

Office of the Press Secretary. 2005. *Joint Statement between President George W. Bush and Prime Minister Manmohan Singh* Office of the Press Secretary, White House, 18 July 2005 [accessed 29 November 2005]. Available from http://www.whitehouse.gov/news/releases/2005/07/20050718-6.html.

Ozga, Deborah. 2000. America the Rogue: The Search for Security through Superiority. *Disarmament Diplomacy* (46).

Perkovich, George. 1999. *India's Nuclear Bomb*. New Delhi: Oxford University Press.

Perkovich, George. 2005. Faulty Promises: The US-India Nuclear Deal. *Policy Outlook*:1-14.

Perkovich, George, Jessica T. Mathews, Joseph Cirincione, Rose Gotemoeller and Jon Wolfsthal. 2005. *Universal Compliance: A Strategy for Nuclear Security* Carnegie Endowment for International Peace, March 2005 [accessed 11 March 2006]. Available from http://www.carnegieendowment.org/files/UC2.FINAL3.pdf.

Rajaraman, R., Zia Mian and A.H. Nayyar. 2004. Nuclear Civil Defence in South Asia: Is It Feasible? *Economic and Political Weekly* 39 (46 and 47):5017-26.

Ramachandran, R. 2000. Thwarted Nuclear Ambitions. *Frontline*, 21 January, 90-93.

Ramana, M. V., Antonette D'Sa and Amulya K. N. Reddy. 2005. Economics of Nuclear Power from Heavy Water Reactors. *Economic and Political Weekly* 40 (17):1763-73.

Ramana, M.V. and Surendra Gadekar. 2003. The Price We Pay: Environmental and Health Impacts of Nuclear Weapons Production and Testing. In *Prisoners of the Nuclear Dream*, edited by M. V. Ramana and C. R. Reddy. Hyderabad: Orient Longman.

Rethinaraj, T.S. Gopi. 1998. ATV: All at Sea Before It Hits the Water. *Jane's Intelligence Review*.31-35.

Ruppe, David. 2005a. *Critics Blast Indo-U.S. Nuke Deal* Nuclear Threat Initiative, 29 November 2005a [accessed 29 November 2005]. Available from http://www.nti.org/d_newswire/issues/2005_11_29.html#09ACED2E.

Ruppe, David. 2005b. *Iran Seeks "Latent" Nuclear Capability, Expert Says* Nuclear Threat Initiative, 25 March 2005 2005b [accessed 30 March 2005]. Available from http://www.nti.org/d_newswire/issues/2005_3_24.html#CF5B39DB.

Sagan, Scott and Kenneth Waltz. 2003. Indian and Pakistani Nuclear Weapons: For Better or for Worse. In *The Spread of Nuclear Weapons: Debate Renewed*, edited by S. Sagan and K. Waltz. New York and London: WW Norton.

Sarabhai, Vikram. 1974. *Science Policy and National Development*. Delhi: Macmillan.

Serrat, Oscar. 2005. *Argentina: Venezuela Sought Nuclear Information* Breitbart.com, 10 October 2005 [accessed 11 March 2006]. Available from http://www.breitbart.com/news/2005/10/10/D8D5IBU82.html.

Singh, Jaswant. 1998. Against Nuclear Apartheid. *Foreign Affairs* 77 (5):41-52.

Srinivasan, G. 2005a. ESCAP Pegs India's Growth Rate at 7.5 pc. *Hindu Business Line*, 27 April.

Srinivasan, M. R. 2005b. New Opportunities for Nuclear Energy. *Hindu*, 2 August.

Srivastava, Sanjiv. 2005. *Indian PM Feels Political Heat* BBC, 26 July 2005 [accessed 11 March 2006]. Available from http://news.bbc.co.uk/go/pr/fr/-/2/hi/south_asia/4715797.stm.

Wilson, Dominic and Roopa Purushothaman. 2003. *Dreaming with the BRICs: The Path to 2050* [Report]. Goldman Sachs, October 2003 [accessed 3 March 2006]. Available from www.gs.com/insight/research/reports/report6.html.

Suggested Readings

Anon. 2005. Left Attacks Joint Statement. *Hindu*, 22 July.

Anon. 2005. *Statement by Prime Minister Dr Manmohan Singh made in Parliament on his recent visit to United States of America* Embassy of India, Washington DC, 29 July 2005 [accessed 14 November 2005]. Available from http://www.indianembassy.org/press_release/2005/July/31.htm.

Anon. 2005. *Transforming India-US Relations: Speech by Indian Foreign Secretary Shyam Saran to CEIP* Carnegie Endowment, 21 December 2005 [accessed 3 January 2006]. Available from http://www.mea.gov.in/.

Anon. 2006. Deal a win-win for India, U.S. *Hindu*, 4 March.

Anon. 2006. *IAEA cheers N-powerment of India* CNN-IBN, 3 March 2006 [accessed 3 March 2006]. Available from http://www.ibnlive.com/article.php?id=6223&section_id=3#.

Anon. 2006. Nuclear separation plan a "surrender" to US: BJP. *Outlook*, 5 March.

Bidwai, Praful. 2005. A Deplorable Nuclear Bargain. *Economic and Political Weekly* 40 (31).

Brinkley, Joel. 2005. U.S. Nuclear Deal With India Criticized by G.O.P. in Congress. *New York Times*, 31 October.

Burns, R.Nicholas. 2005. *Briefing on the Signing of the Global Partnership Agreement Between the United States and India* US State Department, 19 July 2005 [accessed 17 November 2005]. Available from http://www.state.gov/p/us/rm/2005/49831.htm.

Einhorn, Robert. 2005. Undermines US National Interest. *Outlook*, 11 November.

Gahlaut, Seema. 2005. *U.S.-India Nuclear Deal Will Strengthen Nonproliferation* Japanese Institute of Global Communications, 1 September 2005 [accessed 14 December 2005]. Available from http://www.glocom.org/special_topics/asia_rep/20050901_asia_s101/.

Ganguly, Sumit. 2005. *Giving India a pass* Foreign Affairs, August 17 2005 [accessed 1 March 2006]. Available from http://www.foreignaffairs.org/20050817faupdate84577/sumit-ganguly/giving-india-a-pass.html.

Haniffa, Aziz. 2005. *Interview with Stephen Cohen* Rediff, 30 September 2005 [accessed 10 November 2005]. Available from http://specials.rediff.com/news/2005/sep/30inter1.htm.

Kimball, Daryl G. 2005. *A Nonproliferation Reality Check* Economic and Political Weekly, 27 August 2005 [accessed 11 November 2005]. Available from http://www.epw.org.in/showArticles.php?root=2005&leaf=08&filename =9037&filetype=html.

Krepon, Michael. 2005. *"We are breaking rules to accommodate India"* Rediff, 1 September 2005 [accessed 10 November 2005]. Available from http://in.rediff.com/news/2005/sep/01ninter1.htm.

Linzer, Dafna. 2005. Bush Officials Defend India Nuclear Deal. *Washington Post*, July 20, A17.

Linzer, Dafna. 2005. Congress Faults N-deal with India. *Washington Post*, 9 September, A8.

Menon, Kesava and Nirupama Subramaniam. 2005. *Nuclear Issue: "India is a Unique Case"* Hindu, 2005 [accessed 19 November 2005]. Available from http://www.hindu.com/thehindu/nic/mulford.htm.

Mian, Zia and M. V. Ramana. 2005. *Feeding the Nuclear Fire* Economic and Political Weekly, 27 August 2005 [accessed 2005 14 November]. Available from http://www.fpif.org/fpiftxt/659.

Milbank, Dana, and Dafna Linzer. 2005. US, India May Share Nuclear Technology. *Washington Post*, 19 July, A01.

Ministry of External Affairs, India. 2000. *Suo Motu Statement Made in the Parliament on May 9, 2000 by the Minister of External Affairs on the NPT Review Conference* Globalsecurity.com, 9 May 2000 [accessed 28 March 2005]. Available from http://www.globalsecurity.org/wmd/library/news/india/2000/eam-9may.htm.

Nayar, K.P. 2005. No-choice Delhi votes with US. *Telegraph*, 25 September.

Office of the Press Secretary. 2005. *Joint Statement between President George W. Bush and Prime Minister Manmohan Singh* Office of the Press Secretary, White House, 18 July 2005 [accessed 29 November 2005]. Available from http://www.whitehouse.gov/news/releases/2005/07/20050718-6.html.

Perkovich, George. 2005. Faulty Promises: The US-India Nuclear Deal. *Policy Outlook*:1-14.

Potter, William C. 2005. *India and the New Look of U.S. Nonproliferation Policy* Center for Nonproliferation Studies, 25 August 2005 [accessed 8 March 2006]. Available from http://cns.miis.edu/pubs/week/050825.htm.

Ruppe, David. 2005. *Critics Blast Indo-U.S. Nuke Deal* Nuclear Threat Initiative, 29 November 2005 [accessed 29 November 2005]. Available from http://www.nti.org/d_newswire/issues/2005_11_29.html#09ACED2E.

Samanta, Pranab Dhal. 2006. Iran picks on 'objectionable' Indo-US deal. *Indian Express*, 6 March.

Sen, Ashish Kumar. 2005. Now, the Inquest. *Outlook*, 19 December.

Singh, Manmohan. 2005. *Coming to America: PM Singh's Address to Joint Session of US Congress* Indian Express, 20 July 2005 [accessed 22 August 2005]. Available from http://www.indianexpress.com/full_story.php?content_id=74721.

Sudarshan, V. 2006. Fusion Material. *Outlook*, 13 March.

Talbott, Strobe. 2005. *Good Day for India, Bad Day for Non-Proliferation* Yale Global, 21 July 2005 [accessed 17 November 2005]. Available from http://www.brookings.edu/views/articles/talbott/20050721.htm.

Tellis, Ashley. 2005. India: The Only Country Worthy Of Special Treatment. *Outlook*, 11 November.

US Embassy, New Delhi. 2005. *New Framework for the US-India Defense Relationship* US Embassy, India, 28 June 2005 [accessed 8 March 2006]. Available from http://newdelhi.usembassy.gov/ipr062805.html.

Vajpayee, Atal Behari. 2005. *Statement by Shri Atal Bihari Vajpayee on the Joint Statement signed by PM Manmohan Singh and President Bush* Bharatiya Janata Party, 20 July 2005 [accessed 22 August 2005]. Available from http://www.bjp.org/Press/July_2005/july_2005.htm.

Contributors

Invited Participants

Seema Gahlaut

Seema Gahlaut is the Director of the South Asia Program and Senior Research Associate at the Center for International Trade and Security (CITS), University of Georgia. CITS is the foremost non-governmental organization in the world engaged in public policy research and outreach on technology trade and non-proliferation issues. Gahlaut coordinates CITS non-proliferation and export control training programs for the Defence Research and Development Organization (Government of India), Chinese government officials, the China Arms Control and Disarmament Association, and for Indian and Chinese dual-use industry. She regularly briefs government officials in India, the US, and in other European countries on export control issues. As a member of the Export Control Experts Group in the WMD Working Group of the CSCAP (Council and Security Cooperation in the Asia Pacific), Gahlaut follows export control issues in India, Pakistan, Singapore, and other ASEAN states, as well as in the Australia Group and the NSG.

Neil Joeck

Neil Joeck is a Senior Fellow at the Center for Global Security Research at the Lawrence Livermore National Laboratory and an Adjunct Professor of Political Science at the University of California, Berkeley. He served from 2004 to 2005 as Director for Counterproliferation Strategy at the National Security Council, where he was primarily responsible for India and Pakistan proliferation issues. From 2001-2003, he was a member of the Policy Planning Staff at the Department of State, where he was responsible for the India, Pakistan, Afghanistan and nuclear proliferation portfolios. Joeck received a Ph.D. from the University of California, Los Angeles in 1986. His publications include *Maintaining Nuclear Stability in South Asia*, and two edited books: *Arms Control and International Security* and *Strategic Consequences of Nuclear Proliferation in South Asia*. He has contributed articles to *Comparative Strategy*.

S. Paul Kapur

Paul Kapur is an assistant professor of government at Claremont McKenna College in California. He has a Ph.D. from the University of Chicago, and has been a postdoctoral fellow at the Center for International Security and Cooperation at Stanford University since 2004. His research focuses on the strategic effects of nuclear weapons proliferation and on the international security environment in South Asia. His work has appeared in journals such as *Asian Security* and *Security Studies*. The Fall 2005 issue of *International Security* features his article on the causes of war between the nuclear powers India and Pakistan. Kapur's book manuscript, *Dangerous Deterrent: Nuclear Weapons Proliferation and Conflict in South Asia*, is under contract with Stanford University Press.

Ross Neil

Ross Neil began his career in 1994 as an environmental engineer conducting ecosystem and human health risk assessments for the uranium-mining sector. In 1996 he joined the Radiation and Environmental Protection branch of Canada's nuclear regulatory agency, where he built extensive experience in pathways modeling used to determine radiological release limits for nuclear facilities in Canada. In 2001, he joined the Office of International Affairs, Non-proliferation Division as desk officer for South Asia, China and the Middle East, and was a delegate to multilateral nuclear export control groups such as the Zangger Committee and the NSG. In 2004, Neil completed a Masters degree in Political Geography and International Relations with a thesis investigating the private-sector use of satellite remote sensing technology for environment and security applications. In 2005, Neil joined Environment Canada where he currently works as a Senior Policy Advisor in International and Intergovernmental Affairs.

T.V. Paul

T.V. Paul is Professor of International Relations in the Department of Political Science at McGill University in Montreal where he has been teaching since 1991. Paul specializes and teaches courses in international relations, especially international security, international conflict & conflict resolution, regional security and South Asia. He received an M.Phil. from Jawaharlal Nehru University and a Ph.D. in Political Science from the

University of California, Los Angeles. Paul has published eight books, including *India in the World Order: Searching for Major Power Status, Power versus Prudence: Why Nations Forgo Nuclear Weapons and Asymmetric Conflicts: War Initiation by Weaker Powers.* He has published nearly 30 journal articles and book chapters and has lectured at universities and research institutions internationally. He is currently working on two book projects on the nuclear taboo in world politics and on globalization and the national security state.

M.V. Ramana

M.V. Ramana, Ph.D., is currently a faculty member at the Centre for Interdisciplinary Studies in Environment and Development, Institute for Social and Economic Change, Bangalore. He was previously a member of the research staff at Princeton University's Program on Science and Global Security. Ramana studies nuclear weapons and nuclear energy in India and global nuclear disarmament. His work spans both the technical and political aspects of these issues. He is active with the Indian and US peace movements and also serves on the Global Council of Abolition 2000, a network of over 2,000 organizations in more than 90 countries working for the elimination of nuclear weapons. He is the editor of *Prisoners of the Nuclear Dream*, a critique of the Indian nuclear program, and has contributed articles to journals such as *Science and Global Security*.

Ernie Regehr

Ernie Regehr, O.C., LL.D., is Co-Founder and now Senior Policy Advisor of Project Ploughshares, and Adjunct Associate Professor in Peace and Conflict Studies at Conrad Grebel University College, University of Waterloo. His publications on peace and security issues include books, monographs, journal articles, newspaper and magazine articles, conference papers, working papers, and Parliamentary briefs. He has served as an NGO representative and expert advisor on a number of Government of Canada delegations to multilateral disarmament forums, including Review Conferences of the NPT. Among current appointments, he is a Commissioner on the World Council of Churches Commission on International Affairs and on the Board of Directors of the Africa Peace Forum based in Nairobi. In 2003 Regehr was appointed an Officer of the Order of Canada.

Ronald E. Stansfield

Ronald E. Stansfield has been Deputy Director of the Nuclear Non-proliferation and Disarmament Division at FAC since 2003. A graduate of the University of Saskatchewan and Carleton University, he was a member of the Department of Foreign Affairs from 1975-88, during which time his assignments included the Arms Control and Disarmament and Defence Relations Divisions, in addition to postings in South Korea and The Netherlands. In 1988, he was appointed Director of Parliamentary Affairs at the Department of National Defence. He joined the Canadian Nuclear Safety Commission as Advisor on Nuclear Non-proliferation in 1991, remaining with the Commission until the end of 2002. During his time with the Commission, he was also seconded to the IAEA in Vienna from 1998-2001 where he handled non-proliferation and safeguards policy matters.

Simons Centre Participants

Wade L. Huntley

Wade L. Huntley, Ph.D., is Director of the Simons Centre for Disarmament and Non-Proliferation Research at the Liu Institute for Global Issues, UBC. He was previously an Associate Professor at the Hiroshima Peace Institute in Hiroshima, Japan, and has served as Director of the Global Peace and Security Program at the Nautilus Institute for Security and Sustainable Development. His areas of expertise include international security, nuclear non-proliferation and arms control, political relations in the Asia-Pacific region, and political theory. He has published work addressing issues including nuclear weapons developments in East and South Asia, US missile defence ambitions and deterrence policies, the relationship of democracy and peace, and philosophies of science. He received his Ph.D. in political science from the University of California at Berkeley in 1993, where he also received his M.A. in political science in 1985, and his B.A. in economics and political science in 1983.

Karthika Sasikumar

Karthika Sasikumar is a Postdoctoral Fellow at the Simons Centre for Disarmament and Non-Proliferation in 2006. She completed her M.Phil.

at Jawaharlal Nehru University, New Delhi and a Ph.D. in International Relations at Cornell University. Her dissertation explores the interaction between India and the international nuclear non-proliferation order, and its implications for the emerging global counter-terrorism regime. Her major field is International Relations and her research interests are in International Relations theory, international security regimes in nuclear weapons, space, and South Asia. In 2004-05 she was a Pre-doctoral Fellow at the Center for International Security and Cooperation at Stanford University.

Michael Cohen

Michael Cohen is an M.A. student in the Political Science Program at UBC. His research focuses on the intersection of rationalist and constructivist theories, nuclear deterrence, and proliferation.

Lance Noble

Lance Noble is an M.A. student in the political science program at UBC. His research focuses on Chinese foreign policy.

Ariel Zellman

Ariel Zellman is an M.A. student in the political science program at UBC. His research focuses on failed and failing states in Africa and the Middle East, with particular attention to the role played by non-state armed actors in intrastate conflict.

www.ingramcontent.com/pod-product-compliance
Lightning Source LLC
Chambersburg PA
CBHW051449280526
45785CB00003B/1487